7 16 19

Central Heating

Combination Boilers

Fault Finding & Repair

Author:- John Reginald

The author, publishers, contributors and printers do not accept any responsibility of any kind for any injury, death, loss, damage or delay due to error, omission, negligence or however caused resulting from the use of this book. Any and all such liability is discharged. This book is intended as guidance only and is not intended to be exhaustive or definitive and it will be necessary for all parties to exercise skill and judgement when making use of it.

In using this book each and every user by doing so agrees to accept full responsibility for their action and therefore agrees to defend, indemnify and hold harmless the author, publisher, contributors and printers of this book from any and all liabilities arising out of or in connection with such use and irrespective of any negligence on the part of those indemnified. This is a condition of use of this book.

August 2004

Published by Copperjob Publishing

ISBN 0-9546013-1-9

www.combiboilerrepair.co.uk

Acknowledgements

The author expresses his thanks and gratitude for the help extended to him in the compilation of this manual to the following people and organisations:-

Mike Windsor
Ernie McDonald
Clare Southard
Steve Tomaszewski
Trevor Hallett
Diane Witts
Elaine Douthwaite
Nital Shah
Jonathan Collier & Bob Marsh
Shaun Hore
Richard Wilde
Martyn Bridges
Anton
Cookson Electronics (Fernox)
Danfoss Randal Ltd
DifferentDigital Photography
Freezemaster Ltd
Grundfos Pumps Ltd
Honeywell Control Systems Ltd
Martindale Electric Company Ltd
Monument Tools Ltd
On-line Boiler Spares Ltd
Telegan Gas Monitoring
Worcester Bosch Group

Janet Farrage for the assistance, guidance and patience needed in the planning, preparation and production of this book.

Printed by **COLOUR WORKS** Totnes, Devon

www.colour-works.co.uk

Central Heating Combination Boilers Fault Finding & Repair

Overview

Contents

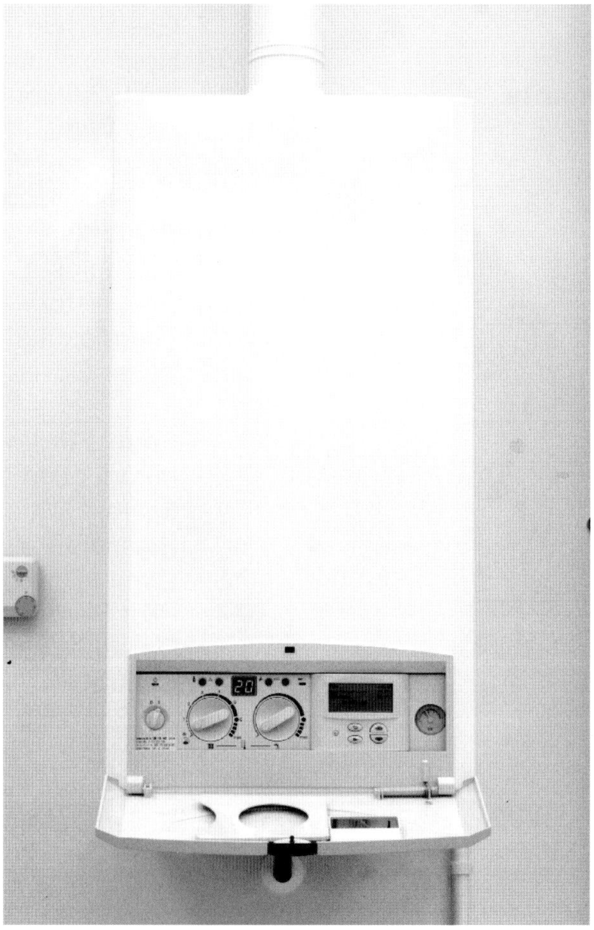

Introduction

To service or fault find on a combi boiler is not about learning one particular model and expecting all others to be just like it. To efficiently service and fault find combi boilers is about understanding the concepts and principles of combi boiler components and how they work, especially in relation to each other. Consider the following:-

- Combi Boilers have been installed in the UK since approximately 1980
- Combi boilers last 15 years or more
- 25 (approximately) major combi boiler manufacturers
- Free standing and wall mounted combi boiler options
- Traditional and balanced flue options
- Standard efficiency and high efficiency (condensing) models
- A range of outputs for each model
- A range of models to suit varying budgets
- Models changing and evolving significantly every 5 years or so

It can be readily seen that there are literally 100's of models and many variations. It is important, therefore, to understand combi boilers generically. For example, if the fan is running but there is no ignition the fault is at or after the fan but before the boiler firing. This is always true. It will be a fault with the fan speed, flue airways, air pressure hose(s) or thermistor(s), air pressure switch or its connections, printed circuit board, gas valve, ignition components, condensate sump. All these components need to be checked unless it is established that the boiler fires in domestic hot water mode but not in central heating mode. Firing in domestic hot water mode eliminates all of the above possible faults but now includes the low limit thermostat, diverter valve and time switch, central heating part of the printed circuit board, room thermostat/programmer, etc *(see Fault Finding – 7)*.

A logical step wise approach is required to quickly track down, isolate and repair the fault, this book will help you achieve this to a professional standard.

Many leading manufacturers offer courses, usually 1 day, on the installation, service, fault finding and repair of their combi boilers. Accompanying these 'hands on' type courses is the manufacturers installation, service, repair and fault finding instructions. The manufacturers course and instructions are invaluable to the engineer…like this book!

This book will give you an understanding of:-

1. Regulations and requirements for installing combi boilers
2. The component parts and their possible faults
3. The role of components in the sequence of operation and shut down
4. Why the controls are used and how they work
5. The correct approach to fault finding and repair
6. How and what to do at service

'Always follow the Manufacturer's Instructions'

A note to enthusiastic DIY/practical householders. You must not use or allow to be used any equipment/appliance you think may be faulty. Have your combi boiler serviced regularly (yearly) by a competent person (CORGI Registered). You must comply with the law.

1.1 What is a Combination Boiler?

Introduction
Boilers which heat hot water for space heating (radiators) and also for domestic hot water direct to the hot water draw off points are known as combination boilers or more commonly combi boilers. Unlike a conventional central heating and domestic hot water system a combi boiler does not store domestic hot water in a separate large domestic hot water cylinder. It heats cold water directly from the rising main.

A combination boiler or 'combi' is a boiler which has a pump, heat exchanger, diverter valves (usually), expansion vessel and controls all inside one unit. Consider a combination boiler to be two appliances housed within one case i.e. a central heating boiler and an instantaneous water heater for the domestic hot water.

It is unlike a conventional heating and hot water system. There's no hot water cylinder, no feed and expansion cistern in the loft and therefore no roof space pipe work and associated insulation. Water for domestic hot water is supplied from the cold mains supply.

A combi saves space and reduces hot water costs, supplying hot water at mains pressure. It provides central heating and domestic hot water. Usually the heat output of combi boilers is governed by the hot water requirements of the property. As this is often more than space heating requirements, modern combi boilers are designed with modulating burners which reduce output to meet the lower space heating demand. Some combi boilers will modulate whilst in domestic hot water mode.

Combination Boiler Types
There are two basic designs of combi boiler:-

Direct combination boiler
Indirect combination boiler

The direct combination domestic hot water pipes run close to the primary heat exchanger within the combustion chamber. The indirect utilises a secondary heat exchanger positioned outside the combustion chamber but within the combi boiler casing.

There are floor and wall combis available with four main categories of combi boilers:-

'Instantaneous' – No internal hot water store.

'Keep Hot' – The water contained in the boiler is kept at temperature to reduce the warm-up time.

A combination storage boiler is a variation of the instantaneous combi and gives greater hot water flow. The stored hot water will give up its heat to provide an improved initial hot water delivery. When the stored heat has been given up to the domestic hot water flow demand the boiler then operates as an instantaneous combi. The hot water flow depends upon the size of the hot water store and does vary.

'Small Store' – The appliance casing contains a small hot water store sufficient for small draw off demands (up to 25 litres) – not enough for a bath.

'Large Store' – The appliance casing contains a large hot water store sufficient for large draw-off demands (up to 50 litres) – bath requirements or multiple draw off points.

Note:- All the above are available as condensing models but it will be approximately 2020 before nearly all installed combi boilers are condensing.

Condensing Boilers
Condensing combi boilers are always more efficient than traditional combi boilers because they extract more heat from the combustion gases. Traditional combi boilers produce a high flue gas temperature between 150°C – 250°C resulting in heat going to waste to the atmosphere. The heat within the flue

gases is referred to as 'latent heat' and much of this is captured within condensing combi boilers by using either a larger heat exchanger or in some instances a secondary heat exchanger *(see Heat Exchangers – 6.13)*. Collecting this otherwise wasted heat increases the efficient of the condensing combi boiler from around 80% to over 90%.

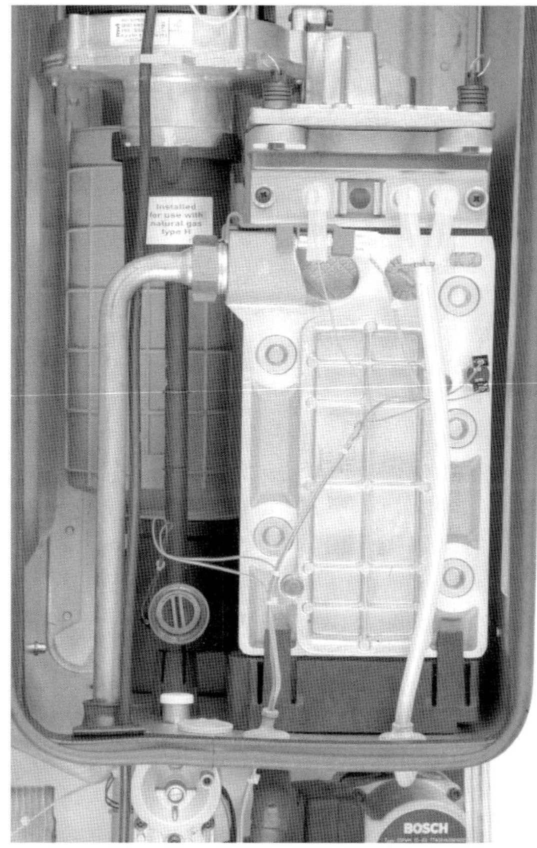

Downward firing condensing combi boiler

This has the effect of cooling the flue gases to 50°C – 60°C allowing the flue system to be made of plastic. It also has the effect of causing a cloud of water vapour to be formed at the flue terminal known as 'pluming.' Therefore special attention must be paid to the siting of the flue terminal to prevent this plume of water vapour causing a nuisance to the inhabitants of the dwelling and neighbouring properties *(see Flues & Terminals – 3.1)*.

The water vapour formed will also condense within the flue forming a gaseous liquid known as 'condensate.' This condensate contains dissolved flue gases and has a pH level of 3.6, similar to tomato juice and will attack some metals, including copper. Natural gas contains about 1.6 litres of water/cubic metre and condensing combi boilers can produce up to 4 litres/hour. This is removed by using a plastic sump often with a siphon action and a plastic condensate drain. *(See Condensate Siphon & Drain – 6.4)*. The flue system must be installed with a fall back to the combi boiler of at least 2.5° so the condensate formed within the flue pipe runs back to the boiler *(see Flues & Terminals – 3.1)*. As the condensate is acidic the heat exchanger is made of aluminium or stainless steel.

Traditional central heating systems use a flow and return temperatures of 82°C and 72°C respectively. Condensing combi boiler central heating systems are normally set at 75°C flow and 55°C return. This 20°C differential must also be set on the radiators by balancing the lockshield valves of each radiator *(see 'Central Heating – Fault Finding & Repair' book)*.

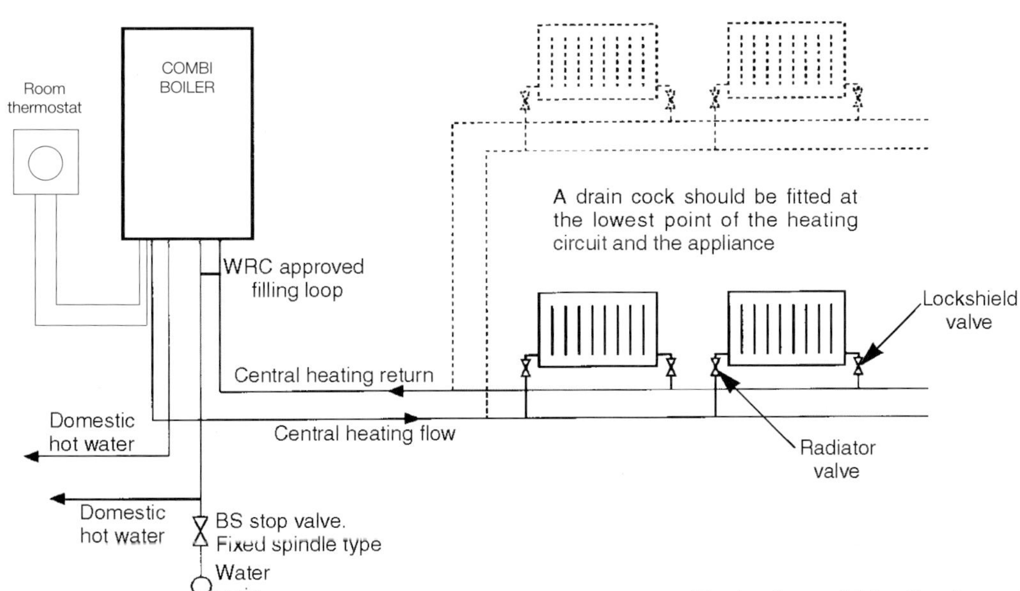

Typical combi boiler layout

1.2 System Layout

Introduction
The traditional central heating system design has evolved into a 2 pipe system supplying many kinds of heat emitters. Panel radiators, towel rails, plinth and skirting heaters, a source of heat energy for domestic hot water and underfloor heating are all well known ways to achieve domestic comfort levels.

The conventional open vented system is often being replaced with a sealed system boiler (*see below*). Many of these sealed systems are now replacing the hot water cylinder and using a sealed system combination boiler.

So, there are regular boilers, system boilers and combination boilers, all can be floor or wall mounted and many are available in condensing form.

Open Vent System (Open to Atmosphere)
An open vented system is a system which includes a separate open safety vent pipe which is open to the atmosphere terminating over the feed and expansion cistern. This cistern also supplies water to the central heating system via the cold feed and expansion pipe.

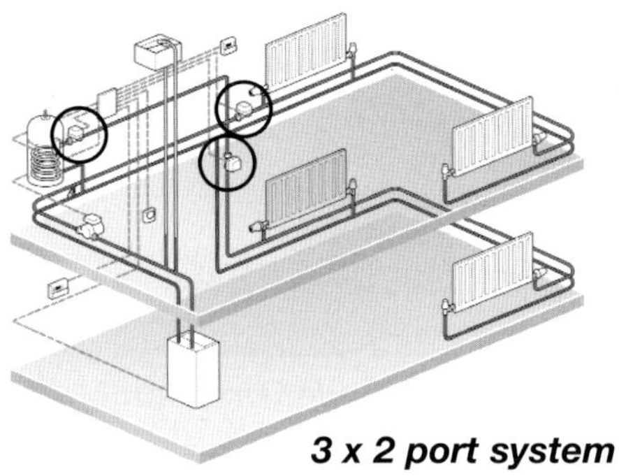

3 x 2 port system

This system, also known as a regular, conventional or traditional system, uses a 'regular' boiler. The boiler provides space heating and is commonly connected to a separate hot water cylinder so providing indirect domestic hot water heating.

The open vent system will include a cold water storage cistern, feed and expansion cistern, associated pipe work and insulation, zone control 2 or 3 port valves, circulating pump, cylinder thermostat, room thermostat/programmer, wiring centre and of course the domestic hot water cylinder.

The open vent system is so called because the primary water, i.e. the central heating system water, is open to the atmosphere through the open safety vent and via the feed and expansion pipe to the cistern.

The open safety vent pipe must ALWAYS terminate over the feed and expansion cistern and NEVER over the cold water storage cistern. The open safety vent pipe is a safety device to allow steam/system water to quickly and safely discharge into the feed and expansion cistern should the boiler controls develop a fault.

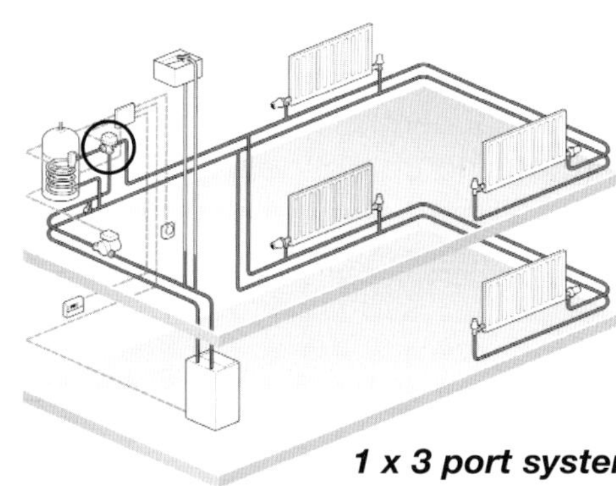

1 x 3 port system

As water expands by approximately 4% when heated, liquids do not compress, the extra volume of expanded water is taken up by the feed and expansion cistern, when the primary water cools it contracts and is 'made-up' from the cistern via the feed and expansion pipe.

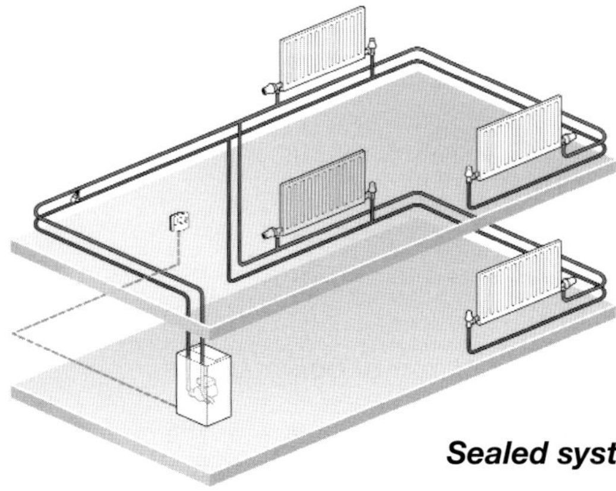

Sealed system

Sealed System (Closed to Atmosphere)
Regular boilers for sealed systems have components such as expansion vessels, pumps, diverter valves etc built in and contained within the boiler casing. They are now referred to as system boilers.

System boilers therefore contain some of the central heating system components found within an open vented system:-

Regular Boiler/Open Vent System	**System Boiler**
Feed & expansion cistern – expansion	Expansion vessel and pressure gauge – *6.7*
Pump positioned within central heating system	Pump positioned within boiler casing – *6.18*
Feed and expansion cistern – cold feed	Filling loop – *6.8*

The sealed system and system boiler installation will require a pressure relief valve to replace the open safety vent pipe. The pressure relief valve is a safety control and must discharge outside the property *(see Pressure Relief Valve – 6.16)*.

Often contained within the casing of a system boiler is an automatic air vent which gives a point for venting the system of trapped air which would normally escape via the open safety vent pipe.

Combination Boilers
The most common type of combination boiler heats domestic hot water on demand without an internal store of pre-heated water.

As with the system boiler containing some of the central heating components the combination boiler contains more of the system components:-

Regular Boiler/Open Vent System	**Combination Boiler/Sealed System**
Feed & expansion cistern – cold feed	Filling loop – *6.8*
Feed & expansion cistern – expansion	Expansion vessel & pressure gauge – *6.7*
Pump positioned within central heating system	Pump positioned within boiler casing – *6.18*
2/3 port zone valve	Diverter valve and flow switch – *6.5*
Hot water cylinder	Not required
Cold water storage cistern	Not required, mains supply to all cold water draw off points
Cylinder thermostat	Water sensor – *6.20*
Wiring Centre	Printed circuit board – *6.6*
Room thermostat	Still required to provide boiler interlock – *6.21*
Open safety vent pipe	Pressure relief valve – *6.16*
Overflow/warning pipe	Pressure relief pipe (discharge pipe) – *6.16*

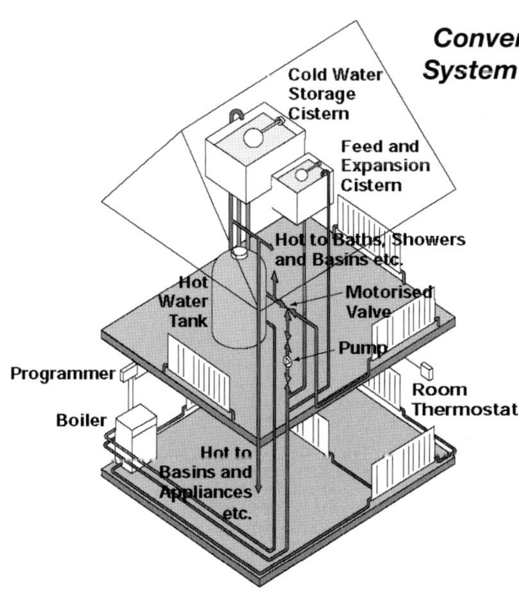

Conventional System Layout

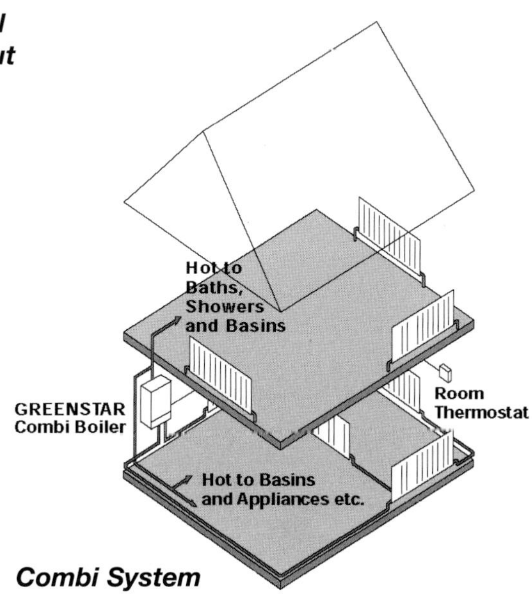

Combi System

Introduction
Having considered the differences between a conventional central heating and domestic hot water system with a sealed system *(see System Layout 1.2)* there are advantages and disadvantages with sealed system combination boilers. These tend to be common to all combination boilers and both the installer and the householder should be aware of these possible advantages and disadvantages.

Advantages

Instantaneous Hot Water – 24 hour/day *(see Dead Legs 2.3)*.

Constant Hot Water – No domestic hot water cylinder to deplete.

Mains Pressure Hot Water – All draw off points at mains pressure – great for showers.

No Contamination – Without a cold water storage cistern in the roof space possible water contamination is eliminated from this source.

'Dry' Roof Space – The need for an elevated feed and expansion cistern and cold water storage cistern are not required (as with a sealed system). This reduces possible freeze problems and also gives more potential for roof space conversion. There is no need for servicing by plumber/engineer in roof space.

Space Saving – A domestic hot water cylinder is not required allowing the 'airing' cupboard to be used for extra storage or conversion to shower cubicle, bathroom extension etc.

No Hot Water Storage – With instantaneous and constant hot water available, costly hot water storage is not required.

Reduced Heat Loss – From domestic hot water cylinder and associated pipe work leading to the feed and expansion cistern. Efficiency should be improved which will reduce wasted energy and will save money.

Less Controls – There is no requirement for a cylinder thermostat, 2 or 3 port zone valves (diverter valves), wiring centre etc. This saves on hard wiring, cost of installation materials and installation time. However, a room thermostat must always be installed to control the central heating system, providing boiler interlock and preventing boiler cycling.

Installation Cost Reduced – Due to dry roof space (no cisterns, valves, pipe work, insulation etc), less controls (cylinder stat, zone valve etc) and labour saving as the installation of combination boiler systems are easier than the conventional layout.

Reduced Corrosion Potential – Sealed systems are closed to the atmosphere. The potential for ingress of oxygen is vastly reduced as 'pumping over' and 'see-sawing' problems are eliminated. Also eliminated is the potential for water loss due to evaporation from the open safety vent pipe and the feed and expansion cistern which require automatic topping up of the feed and expansion cistern. This topping up introduces fresh oxygenated water into the system which now contributes to more corrosion and scale build up.

Disadvantages

No Stored Cold or Hot Water – In the event of a mains water failure the supply of cold and hot water immediately ceases to all draw off points within the property. Nowadays this rarely happens. When there is a break in the mains water supply it is usually repaired quickly and often without the householder ever knowing of a mains water shut down.

Central Heating Cooling – Combination boilers are set as 'hot water priority.' Excessive domestic hot water draw off times prevents the central heating required temperatures being achieved. Should the central heating room thermostat call for heat (demand) when hot water is being drawn off, the central heating demand will not be met until the hot water draw off demand ceases. Only then will the diverter valve move and allow the central heating call for heat to be satisfied. In the event of long excessive hot water draw off times radiators will cool down. This can be a problem in large family households.

Topping Up – All sealed system central heating boilers need occasional 'topping up' using the filling loop to replace water loss from very small minor leaks and from radiator venting etc. The householder must be made aware of this necessary activity with all system boilers or combination boilers using a sealed system installation layout *(See Filling Loops – 6.8)*.

Considerations

Insufficient Hot Water Flow Rates – Many older combination boilers only supply hot water flow rates up to 10 litres/minute which is much less than a conventional domestic hot water cylinder can provide. If a hot water draw off point is opened whilst another is being used one or both will suffer a drastically reduced flow rate. Modern combination boilers can deliver 15+ litres/minute so reduced flow rates are now much less of a problem *(see Waterflow – Pressure – 2.2)*.

More Expensive – As there are more components within a combination boiler they are more expensive than a comparable open vented system boiler installation. However, a full combination boiler installation is always less expensive than a traditional boiler with domestic hot water cylinder, feed and expansion cisterns, zone valves, cylinder thermostats etc.

Appliance Size – Combination boilers are always bigger and require more wall space than its equivalent rated regular central heating boiler. The overall dimensions and clearance tolerances of most modern combination boilers are now installer and 'kitchen friendly.' Manufacturers have recognised the need to reduce the overall size of appliances and to keep the minimum clearances as small as possible. This allows most combination boilers to be fitted/housed within a standard size kitchen unit or cupboard.

Appliance Positioning – A sealed central heating system is normally incorporated as part of a combination boiler so a suitable position needs to be located for the outlet of the pressure relief discharge pipe. It must terminate so that it discharges safely in a visible position i.e. any water discharged will not cause injury as it will be hot and it must be positioned in such a place that the householder will recognise it discharging. As most combination boilers are of the balanced flue type this positioning of the pressure relief discharge pipe is usually not a problem but is a consideration.

Gas Pipe Size – Large (35 kW and above) gas combi boilers place a big demand on the gas supply pipe work and could starve other appliances of gas if this pipe work is incorrectly sized *(see Gas Pipe Sizing – 2.1)*.

No Airing Cupboard – The loss of an airing cupboard can be regretted so a small radiator can be fitted within the cupboard (controlled by thermostatic radiator valve). Unfortunately, heat for the airing cupboard is only available when the central heating is 'on.'

1.4 Operating Sequence

Introduction

When there is a call for central heating or hot water the combi boiler will go through a series of events to meet that demand. When the demand is satisfied the combi boiler will cycle through another series of events and await a further call for heat. All combi boilers do this. As there are no exceptions the fault finding, repair and servicing of combi boilers is simple and straightforward. Not quite, it is if you know exactly what you are doing and you know the sequence of operation for every combi boiler. Luckily, the basic operating sequence is the same for many. There are 'variations on a theme' but if you can recognise the control circuitry and the major components installed the same procedural logic can be applied to all.

Operating Sequence

Fan flued, room sealed combi boilers operating sequence will always be the same and will follow 5 levels:-

Standby – Fan – Ignition – Ignition Proving – Main Burner

There are a series of steps that must occur before the next level is reached, see example below, only when the following is proved will the combi boiler move from

Standby to Fan

- There is an electrical supply
- The electric supply is correct
- The fuse is O.K.
- The overheat thermostat/thermistor is O.K.
- Thermostat/thermistor wiring O.K.
- The low system water pressure is O.K.
- Electronic board O.K.

Fan to Ignition

- The fan is operating correctly
- The air pressure switch is O.K.
- The pump is O.K.
- The hot water flow switch is O.K.
- Electronic board O.K.

Ignition to Ignition Proving

- The fan is working effectively
- Electrode leads O.K.
- Ignition electrode O.K.
- Spark generator O.K.
- Sparking O.K.
- Pilot Solenoid O.K.
- Gas supply O.K.
- Electronic board O.K.

Ignition Proving to Main Burner

- Electrical supply O.K.
- Flame rectification probe O.K.
- Flame rectification lead O.K.
- Electronic board O.K.

Main Burner

- Gas supply O.K.
- Gas valve is O.K.

Operating Sequence 1.4

Combination boilers do vary in the exact series of events but importantly it will not progress from one level to another if any of the feedback signals are absent/incorrect. By working within these 5 levels, understanding the general operating sequence, recognising the components within the combi boiler, having the correct tools/instruments and knowing how to use them you will soon develop your own effective fault logic.

The following is an example of a typical operation sequence and lists steps that should be considered as you follow your fault logic.

Combi boiler on	If not check:-	On/Off switch is on Pressure and gauge O.K. 240 Volts at combi boiler main terminal Fuse(s) Thermostat/Thermistor and wiring
Fan & pump running	If not check:-	240 Volts at fan & pump Fan or pump are not jammed
Gas valve lifting	If not check:-	Air pressure switch or thermistor Hoses are not split and good connections to hose ports Electrical terminal connections are O.K. 24Volts at gas valve solenoid Solenoid O.K.
Sparking O.K.	If not check:-	Ignition leads for integrity Lead terminals are sound Spark gap correct Correct voltage at spark generator
Burner Lights	If not check:-	S/B Gas on Electrode lead is not earthing out The electrode is not cracked/pitted Injector is not faulty Gas rate is correct

The fault logic can now be continued if say a hot water or central heating demand is not met.

Always try and obtain a copy of the Manufacturers Instructions

Installation Considerations

Contents

Introduction

Before installing a combi boiler it is ALWAYS necessary to consider ALL of the issues that concern the householder and the installer. The initial site survey should include discussions with the householder, as their requirements and preferences must be taken into account when the installer makes final decisions. It is equally important that the householder understands exactly what the installation involves and the implications of owning a combi boiler.

Installation Considerations

Listed below is a checklist of points for the installer to consider and in many cases discuss with the householder:-

Customer Requirements

Lifestyle – A 40 kW combi boiler with domestic hot water flow rate of 16 Litres/min may suit a family dwelling with up to 4 bedrooms and 2 bathrooms. However, with such a large property and say 3 teenage children combined with en suite shower it may necessitate a combi boiler with hot water storage facility or a combi boiler combined with an indirect domestic hot water cylinder.
Property – A large 4/5 bedroomed house with several bathrooms and being run as a guest house would require stored water for the guests with perhaps the combi boiler serving the kitchen and private side of the property *(see What is a Combination Boiler? – 1.1)*.
Hot Water Flow Rates – Establish what domestic hot water flow rates the customer expects and perhaps demonstrate flow rates by setting up cold water and draw off points using a weir gauge *(see Water Flow/Pressure – 2.2)*.
Cold Water Storage – If the customer requires cold water storage the cold water storage cistern must remain. Ensure it has a well fitting lid, is correctly insulated and complies with cold water storage regulations *(see 'Central Heating – Fault Finding & Repair' book)*.
Topping Up – Ensure the customer is aware that the central heating system will require 'topping up' from time to time and explain what this actually involves. Explain the use of the filling loop and pressure gauge and the implications of venting radiators etc *(see Filling Loops – 6.8)*.

Cold Water Supply

Mains Supply Pressure – Using a pressure gauge establish there is at least 1.5 bar pressure at the mains fed cold water draw off point, usually at the kitchen sinks. Remember that this may vary at different times during the day. If the pressure is greater than 10 bar explain that a pressure reducing valve must be fitted to protect the combi boiler components and system, pipe work and fittings *(see Water Flow/Pressure – 2.2)*.
Mains Cold Water Flow Rate – Using a weir gauge establish there is at least 20 Litres/min flow rate at the mains fed cold water draw off point, usually at the kitchen sink. Remember that this may vary at different times during the day *(see Water Flow/Pressure – 2.2)*.
Scaling – In hard water areas it may be necessary to fit a scale reducer to help protect the main heat exchanger and the calorifier.

Existing Pipe Work & Fittings

Active Dead Legs – Establish the lengths of active dead legs and explain to the customer the implications. Active dead legs of more than 12 metres are generally unacceptable *(see Dead Legs – 2.3)*.
Passive Dead Legs – Carefully inspect the existing installation and check for existing passive dead legs and consider removing. Ensure that no passive dead legs will be created at install *(see Dead Legs – 2.3)*.
Taps – Are all the cold and hot water taps in the dwelling suitable for mains pressure? If not, they may fail or constantly drip, so consider upgrading. Check the W.C. float operated valve is suitable for mains pressure or inform customer a replacement must be fitted.
Water Regulations – When all the existing is mains fed ensure that Water Regulations are not contravened *(see Water Regulations – 3.3)*.

2.0 Installation Considerations

Combi Boiler Position

Clearances – Consider possible combi boiler positions and ensure there will be sufficient clearances to the sides, top and bottom and to the front of the combi boiler. Always follow the manufacturers recommendations *(see Clearances & Ventilation – 3.1)*.

Flue Route – The flue route and terminal must be carefully considered. Also, due consideration must be given to the 'pluming' effect of condensing boilers. Always follow the manufactures recommendations *(see Flues & Terminals – 3.1)*.

Ventilation – Does the selected combi boiler require additional ventilation. Always follow the manufacturers recommendations *(see Ventilation & Cooling – 3.2)*.

Gas Supply

Gas Pipe Sizing – The gas supply pipe must be correctly sized for the suggested combi boiler rating.

Other Gas Appliances – Establish what gas appliances are already installed within the dwelling and what, if any, gas appliances may be installed and where *(see Gas Pipe Sizing – 2.1)*.

Gas Meter – If the total maximum gas rate of all the gas appliances installed or to be installed exceeds 6.0 m^3/hr will the gas meter be adequate? Most domestic meters are rated at a maximum of 6.0 m^3/hr *(see Gas Pipe Sizing – 2.1)*.

Electrical Supply

Mains Bonding – Establish that the present gas and water installation pipe work is mains bonded *(see Electrical/Wiring – 3.4)*.

Supplementary Bonding – Establish that the present installation in supplementary bonded *(see Electrical/Wiring – 3.4)*.

Combi Boiler External Wiring – Establish the position and suitable of electrical connection to power the proposed combi boiler *(see Electrical/Wiring – 3.4)*.

Room Thermostat – If the present system does not have a room thermostat to provide boiler interlock, one must be fitted and the reason why must be explained to the householder. Consider the cable run from the combi boiler to the room stat and available positions *(see 'Central Heating – Fault Finding & Repair' book)*.

Introduction

When a gas supply is required within a domestic property a service pipe is run to the building. All gas supplied to the property will be supplied, for billing purposes, via a meter. The gas meter may be located in a garage or inside the dwelling or now more usually contained within an external meter box. External meter boxes provide the supplier with easy access to obtain meter readings.

There are 3 main types of external boxes used:-

Sunken Meter Box – Partially submerged in the ground with lift up hinged lid.
Surface Mounted Meter Box – Wall mounted with hinged fold down access or side hinged door.
Semi-Concealed Meter Box – Partially set into wall with side hinged door.

To gain access to the meter contained within the box a special key is often required of which there are several types. The householder MUST be left a key to gain access to operate the emergency/meter control valve.

The supply gas pressure through the service pipe may be up to 75 mbar but is reduced to a supply pressure of 21 mbar ± 1 mbar. This reduction in pressure is required to allow domestic gas appliances to be used and is achieved by the use of a meter governor, now referred to as a regulator, installed immediately before the meter. The regulator automatically adjusts to provide an adequate supply of gas to the meter at a constant pressure of 21 mbar ± 1 mbar. The regulator is set by the gas supplier, is sealed and must NOT be adjusted. Most often the emergency/meter control valve is installed on the service pipe side of the meter *(see Gas Safety – 4.1)*.

All pipe work up to the meter, the meter control valve, the constant pressure governor (regulator) and the meter is the responsibility of the gas supplier. Pipe work leaving the meter to the property is the responsibility of the householder. Therefore, the gas pipe work is the responsibility of the installer and all gas and building regulations must be adhered to.

Pipe work passing through joists etc. are covered by building regulations, however, it is a gas regulation that no gas pipe will be installed in a cavity wall except when passing from one side to the other by the most direct route. When installed through a wall the gas pipe must run through a sleeve (e.g. 15mm pipe through a 22mm pipe sleeve, 22mm through a 28mm sleeve etc). The sleeve is to be sealed (use mastic sealant) at one end only, usually the internal wall, to prevent the passage of gas, to allow for movement and to prevent mortar corroding the gas pipe.

Gas Pipe Sizing

Gas pipe work and fittings must be installed and tested in accordance with gas safety regulations *(see Gas Safety – 4.1)* and the manufacturers instructions.

For the customer, undersized gas pipe work will result in poor combi boiler performance i.e. reduced hot water temperature/poor flow rates at desired temperatures.

Undersized gas pipe work can create a dangerous situation. If the gas pressure is insufficient at the burner i.e. the combi boiler is under gassed, insufficient primary air will be drawn in. This incorrect gas/air ratio may result in explosive ignition, soot formation and deposition, incomplete combustion which may produce carbon monoxide, burner flames extinguishing etc. It is possible for light back into pipe work to occur which may lead to an explosion. This constitutes an 'Immediately Dangerous' situation. It is essential that the maximum pressure drop between the meter and the furthest appliance, under maximum flow conditions, does not exceed 1 mbar.

When installing a combi boiler the gas pipe work must be designed to allow for an adequate flow of gas to each appliance and that manufacturers stated working (operating) pressures are achieved.

2.1 Gas Pipe Sizing

If a 25 kW system boiler is replaced with a 30 kW combi boiler and undersized pipe work gives rise to a 3 mbar reduction from the 20 mbar dynamic pressure, the combi boiler output will be reduced by 5 kW! This will reduce ΔT water temperatures by about 8°C! That's the difference between a cool bath, say 32°C and a comfortably hot bath at say 40°C *(see Hot Water Safety 4.3)*.

Preparation and Planning
When installing a combi boiler all the gas pipe work from the meter to each appliance must be calculated. These calculations should be based on the maximum gas rate of each appliance. This information can be obtained from either the manufacturer's instructions or from the data badge/plate on each appliance.

Consider and discuss the proposed pipe work route with the customer and discuss any possible future extensions/appliances. Make a drawing of the proposed pipe work noting the route, length of pipe runs and fittings to be used. In summary, the following needs to be established:-

Maximum gas rate of each appliance
Route of gas pipe work
Length of each section of gas pipe work
Type and number of fittings to be used

Note:- When establishing the maximum gas rate of the combi boiler DO NOT add the central heating and domestic hot water rates together. Use the maximum rate of the domestic hot water.

An allowance factor for the type and number of fittings (including bends) must be added into the calculation as they offer additional resistance to the gas flow. Each fitting and bend has an 'equivalent length.' For pipe work up to 28mm use these values:-

0.5 metres for each elbow
0.5 metres for each tee
0.3 metres for each bend

The chart below shows discharge rates in m³/hr for straight horizontal copper tube and can be used for calculations of gas pipe sizing:-

	Length of pipe in metres									
Pipe Diameter	3	6	9	12	15	20	25	30	40	50
mm	**Gas discharge m³/hr**									
8	0.52	0.26	0.17	0.13	-	-	-	-	-	-
10	0.86	0.57	0.5	0.37	0.3	0.22	0.18	0.15	-	-
12	1.5	1.0	0.85	0.82	0.69	0.52	0.41	0.34	0.26	0.2
15	2.9	1.9	1.5	1.3	1.1	0.95	0.92	0.88	0.66	0.52
22	8.7	5.8	4.6	3.9	3.4	2.9	2.5	2.3	1.9	1.7
28	18	12	9.4	8	7	5.9	5.2	4.7	3.9	3.5

It can be seen from this chart that a 24 kW combi boiler with a gas rate of 2.8 m³/hr, 2 metres of pipe length from the gas meter and 4 elbows (total equivalent length 4 metres) would need a minimum of 22mm pipe.

A 35 kW combi boiler with a gas rate of 4.1 m³/hr, 8 metres of pipe length from the gas meter and 6 elbows (total equivalent length 11 metres) would require 28mm pipe.

Both the above examples assume there is no other gas appliance in the property and yet a minimum pipe size of 22mm is required. Therefore, a rule of thumb is to always use 22mm minimum diameter pipe work to any size combi boiler.

Before we look at a worked example consider the customer requiring a 35 kW combi boiler with a gas rate of 4.1 m³/hr. Already installed is a 7 kW gas fire with a gas rate of 0.65 m³/hr and a 15 kW range cooker with a gas rate of 1.35 m³/hr. This is a total of 6.1m³/hr. The gas meter installed in most properties is designed to pass a maximum of 6 m³/hr. Therefore, the installation of a 35 kW combi boiler with that gas rate installed with the other appliances is unacceptable. However, gas flow rates do vary with similar rated combi boilers so always carefully check the manufacturers instructions.

Calculation of Gas Pipe Size

When just one gas appliance is installed the discharge rate chart above can be used as is. However, in the following example there are 3 sections i.e. A to B, B to D and the third section is the final section to each appliance; B to C, D to E and D to F. Therefore, the equivalent length MUST be multiplied by 3 to ensure that the required gas rate is available to each appliance for the given pressure loss.

As the chart is based on 1 mbar pressure loss this is divided by the 3 sections to give 0.33 mbar. Example:-

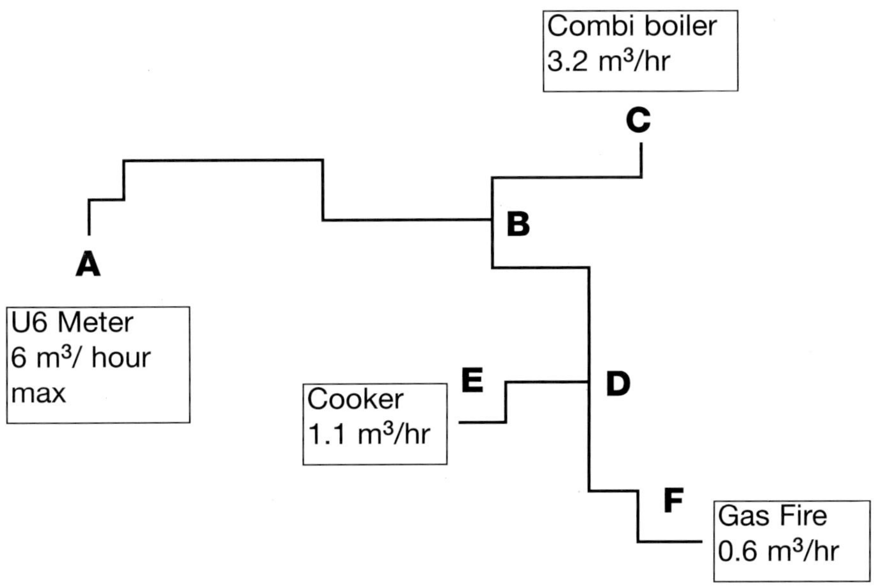

Length of pipe work in each section fitting in each section:-

A to B = 5M	A to B = 5 elbows and 1 tee
B to C = 3M	B to C = 2 elbows
B to D = 1.5M	B to D = 2 elbows and 1 tee
D to E = 4M	D to E = 2 elbows
D to F = 3.5M	D to F = 3 elbows

Pipe sizing calculation – example.

Pipe Section	Gas Rate (m³)	Actual Length (M)	Pipe Fittings Type	Allowance Factor (M)	Equivalent Length (M)	Sections x 3	Pipe Size (mm)
A to B	4.9	5M	5 elbows + 1 tee	3M	8M	24M	28
B to C	3.2	3M	2 elbows	1M	4M	12M	22
B to D	1.7	1.5M	2 elbows + 1 tee	1.5M	3M	9M	22
D to E	1.1	4M	2 elbows	1M	5M	15M	15
D to F	0.6	3.5M	3 elbows	1.5M	5M	15M	12

In the above example when installing a combi boiler the installer is primarily concerned with the gas pipe sizing of A to B and B to C. However, this would assume that the gas pipe work to other appliances is correct!

Introduction

Historically combination boilers have only been suitable for households in which the demand for domestic hot water is low such as 1 bathroom properties. Modern combination boilers, especially condensing combination boilers, have a flow rate of up to 20 litres of water/minute at 40°C. The volume of hot water available will be limited by the flow rate of mains cold water supply and the speed at which it can be heated.

Cold Supply

The mains cold water supply must be of sufficient flow and pressure to allow hot water draw off point(s) to be used at the required rate and temperature and for cold water draw off point(s) to be used at the same time, without adversely affecting the flow at any point.

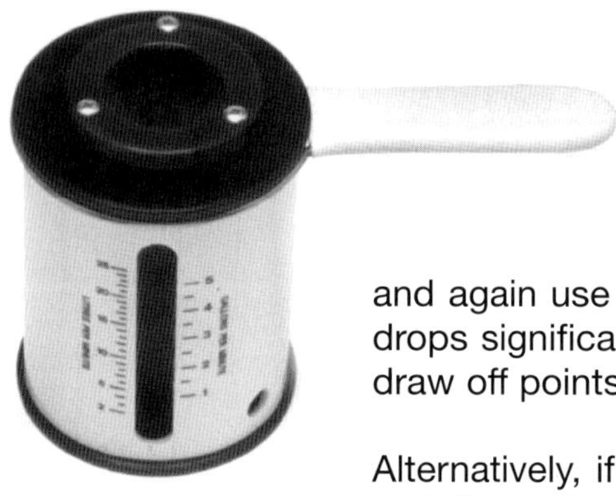

Weir gauge

The flow rate of cold water can be assessed using a weir gauge at the kitchen mains fed draw off point. A flow rate of 20 litres/minute would seem to be adequate but now turn on another cold water draw off point, which must be connected to the mains supply and not to the cold water storage cistern e.g. outside tap, flushing the toilet etc. and again use the weir gauge at the kitchen draw off point. If the flow rate drops significantly say to 10 litres/minute, the flow rates at the hot and cold draw off points may be unsatisfactory if a combination boiler is installed.

Alternatively, if the only mains supplied draw off point is at the kitchen sink and the rest of the property (including the domestic hot water cylinder) is supplied by the cold water storage cistern the following method can be used to assess flow rates. Open the cold draw off point to the bath and flush the toilet. This will now lower the water level in the cold water storage cistern and the float operated valve will open. This will place a demand on the mains supply. Use the weir gauge at the kitchen cold water draw off point and record the result. If 15 litre/minute or more is recorded then a combination boiler and the hot and cold draw off points should operate satisfactorily.

Remember that mains supply flow rate will vary according to the demand placed on it by all the users, domestic and commercial in that area. The highest demand is between 0730 and 0900 hours (high usage of bathrooms and toilets) and the flow rate is usually at the daytime low between 1300 and 1500 hours.

The pressure of the rising main is also important to facilitate the optimum domestic hot water performance of the combi boiler. Attach a pressure gauge to the kitchen cold water draw off point and record the pressure. Too low a reading and the combi boiler will not perform at its best or may not work at all in domestic hot water mode. Too high a pressure may damage and cause leaks in pipe work fittings and valves and cause dripping at draw off points etc.

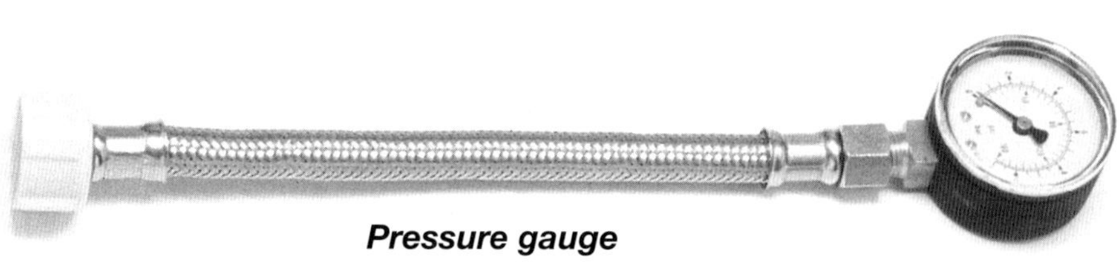

Pressure gauge

A pressure reading of less than 0.5 bar is generally unsatisfactory. If the pressure is greater than 10 bar a pressure reducing valve should be fitted to the mains water side of the installation.

To allow effective commissioning and 'top-up' of the central heating system and to provide for satisfactory performance of the domestic hot water a minimum pressure of 1.5 to 2 bar is required. In summary:-

Mains Water Flow Rate – 20 litres/min minimum
Mains Pressure – 1.5 bar min and 10 bar maximum

Hot Supply

Manufacturers generally specify domestic hot water flow rates as a flow rate of litres/minute at a temperature rise of 30°C or 35°C. This will be a temperature rise on top of the temperature of the incoming mains supply water.

e.g. A boiler is specified as having a domestic hot water flow rate of 15 litres/minute at 35°C ΔT. If the incoming cold water is at 11°C and is raised by 35°C, 15 litres of water per minute at 46°C will be delivered at a hot water draw off point. This will accommodate 2 hot water taps or run a bath in 5 – 7 minutes or give an excellent power shower (always use a thermostatic shower mixer). Combination boilers are designed to raise the water to a maximum temperature of around 57°C (± 3°C) but the flow rate is reduced.

When comparing combination boiler flow rates always ensure that the temperature rise quoted are the same. These temperature rises are usually shown as 30°C ΔT or 35°C ΔT.

Flow Rates

Manufacturers will always specify domestic hot water flow rates (they set the flow restrictor accordingly) and temperature rises but in the absence of information, flow rates can be calculated. This calculation is not difficult so read on:-

A unit of energy is called a Joule.
To raise 1 gram of water by 1°C requires 4.182 Joules. We will take 4.2 as the value.
4.2 is a fixed value and is referred to as 'the specific heat capacity of water.'
If the heat source is limited by the fuel being burnt and the capacity of the combi boiler, there are no variables. Just a question of putting the right values in the right section of the equation.

$$\frac{\text{Boiler Output kW (Kilowatts) x convert minutes to seconds}}{\text{Specific heat capacity of water x Delta T (temperature rise)°C}}$$

Kilowatts (kW) – combi boiler output
Convert minutes to seconds – always 60
Specific heat capacity of water – always 4.2
Delta T – temperature rise °C

If the incoming supply is 10°C and the hot water at the draw off points is to be 45°C, Delta T will be 45°C – 10°C = 35°C. A 24 kW combi boiler would have a flow rate of:-

$$\frac{\text{kW (Kilowatts) x minutes to seconds}}{\text{Specific heat capacity of water x } \Delta\text{T (temp. rise)°C}} = \frac{24 \times 60}{4.2 \times 35} = \frac{1440}{147} = 9.8 \text{ litres/minute}$$

Combination boiler flow rates, temperature rises and boiler outputs may be calculated using the following formulas:-

To calculate the **Water Flow Rate** for a specific temperature rise and known boiler output:-

$$\text{Flow rate (litre/minute)} = \frac{\text{boiler output (kW) x 60}}{\text{4.2 x temperature rise (°C)}}$$

To calculate the **Temperature Rise** for a specific flow rate and known boiler output:-

$$\text{Temperature rise (°C)} = \frac{\text{boiler output (kW) x 6.0}}{\text{4.2 x flow rate (litres/minute)}}$$

2.2 Water Flow/Pressure

To calculate the **Boiler Output** required to provide a specific temperature rise at a specific flow rate:-

$$\text{Boiler output (kW)} = \frac{\text{flow rate (litres/minute) x 4.2 x temperature rise (°C)}}{60}$$

As a guide the following domestic hot water flow rates at 35°C ΔT can be expected from the boiler rating (kW). Manufacturers instructions and specifications always prevail.

	24 kW	28 kW	30 kW	35 kW	40 kW
Domestic hot water gallons	2.1	2.5	2.8	3.2	3.5
Flow rate @ 35°C ΔT litres	9.5	11.5	12.5	14.5	16.5

Introduction

Often overlooked and yet the cause of many post installation problems, there are two types of dead legs:-

Passive Dead Leg

Active Dead Leg

Passive Dead Leg – Where a length of pipe is capped off air is trapped and when a draw off point is opened and closed the air compresses causing a delay in the equilibrium of pressure and possible pipe 'knocking.' With combination boilers, this fault is often created by the installer when the domestic hot water cylinder is removed and the installer caps off the hot water draw off point close to the cylinder. This causes the above fault and causes the combi boiler to remain alight for longer than it should whilst the pressure equalises. The installer MUST remove all unnecessary pipe work and not create a passive dead leg.

Active Dead Leg – The run of pipe from the combi boiler to the draw off point is an active dead leg. When a hot water draw off point is opened the warm/cold water contained within the pipe work between the draw off point and the combi boiler is allowed to run to waste. An excessive length of active dead leg causes unnecessary delay for the user and wastes water. When the draw off point is closed there is a delay in water pressure equilibrium causing the combi boiler to remain alight for longer than necessary. It is also inefficient as the water contained within the dead leg has been heated but not used.

Active Dead Leg

The effective length of a pipe run is the actual length of the pipe plus the equivalent pipe length attributable to copper fittings which cause flow reduction due to increased friction. This frictional loss through copper fittings may be calculated as per the chart below:-

Equivalent copper pipe length, due to frictional loss caused by copper fittings

Pipe Size (mm)	Elbow (M)	Tee (M)	Stopcock (M)	Check Valve (M)
15	0.5	0.6	4.0	2.5
22	0.8	1.0	7.0	4.3
28	1.0	1.5	10.0	5.6

e.g. Actual run of 15mm pipe work to draw off point	=	15M
10 elbows @ 0.5M	=	5M
2 tee @ 0.6	=	1.2M
Total		21.2M

The time taken for hot water to leave the combi boiler and discharge at a draw off point can be calculated by using the following chart:-

Pipe Length v Time Delay for Delivery of Hot Water

Pipe Size	Flow Rate	Maximum Pipe Length for Hot Water to Reach Outlet			
mm		60 seconds	30 seconds	20 seconds	10 seconds
15	6 litres/min	45.0M	22.5M	15.0M	7.5M
15	10 litres/min	75.0M	37.5M	25.0M	12.5M

It can be seen that from the above worked example that a combi boiler providing a flow rate of 10 litre/min at the draw off point will take approximately 20 seconds (plus heat up time) to reach temperatures at the draw off point.

This time lag may be reduced to a more acceptable level by the planned pipe runs being reconsidered and a more direct route designed with less copper fittings and bends.
As the combi boiler heated domestic hot water flows through the pipe work it will loose heat through the pipe walls to the surrounding air causing unnecessary background heating and effectively reducing efficiency. There are recommended maximum runs of un-insulated hot water distribution copper pipe as per the chart below.

Maximum Length of Un-insulated Hot Water Distribution Pipe

Pipe Size(mm)	Length (metres)
15	12
22	8

Again, in the worked example the overall length is greater than 12 metres therefore the domestic hot water pipe must be insulated. This will ensure the hot water discharges at maximum possible design temperature *(see Hot Water Safety 4.3)* increasing efficiency and providing increased customer satisfaction.

Eliminate any possible passive dead leg and design the hot water pipe work system so active dead leg runs are kept to the minimum practical length. Use as few fittings and bends as possible and insulate whenever and wherever possible e.g. in cupboards, under/through/behind kitchen units, under floor boards, in the roof space etc.

Introduction
Corrosion is the process of wasting away and with metals it is the chemical reaction to its surface that may cause it to corrode and eventually fail to do its job. Reduced efficiency levels, leaks and breakdowns are all evidence of corrosion taking place within the central heating and/or domestic hot water systems.

Atmospheric Corrosion
Atmospheric corrosion or oxidation occurs when moisture and gases in the air, more accurately oxygen, mix with the surface of say iron, causing it to form iron oxide – rust. This rust moves away exposing fresh metal which in turn rusts.

$$Fe + O_2 = FeO_2$$ i.e. ferrous oxide or rust

Water contains naturally dissolved oxygen which will cause the mild steel (iron) in central heating systems to corrode by forming oxides. However, once heated the water gives up its oxygen and provided no more air is allowed to enter the system the corrosion process will be significantly slowed down. Under ideal situations the corrosion will be completely inhibited *(see Flushing and Protecting – 2.5)*.

Electrolytic Corrosion
Central Heating systems are made up of various types of metals, copper (pipe work), iron (mild steel radiators) cast iron (combi boiler heat exchanger) brass (valves and other plumbing fittings) aluminium (condensing combi boiler heat exchanger) etc and all are connected by the circulating water.

The central heating water acts as an electrolyte. This water carries electrically charged particles from an anode to a cathode. One metal will act as the anode and another as the cathode and the rate of corrosion will depend on a) the water (what it contains) and b) where each metal appears in the electromotive series – *see below*.

Electromotive Series
Copper
Tin
Lead
Nickel
Iron
Zinc
Aluminium
Magnesium

All that is needed to be known here is those metals higher on the list will gradually destroy those lower down e.g. Copper will destroy (corrode) Iron. The further away the metals are in the list the more aggressive is the corrosion.

Manufacturers of domestic hot water cylinders often install a rod of magnesium into the vessel to act as a sacrificial anode. In some properties a block of aluminium is placed into the cold water storage cistern which will also act as the least noble metal in the system and corrode before system components.

Dezincification of Brass
Brass, composed of copper and zinc, is also used in central heating systems and suffers from a condition when the zinc content of the brass is destroyed leaving it porous and brittle. This dezincification is a type of accelerated electrolytic corrosion. Where does the zinc go? It Is converted into a basic carbonate which now contributes to a blockage in the pipe work, radiators, combi boiler, valves, pumps etc.

Chlorine

The mains water supply contains added chemicals such as chlorine and sometimes fluorine. Add these now chlorides and fluorides to the naturally occurring soil sulphates and carbonates (depending on geographical location) and there is further potential for corrosion.

Installation of Central Heating System

During installation the system will have various corrosive compounds introduced into it, substances such as plumbers flux, (used at soldered joints) dust, copper swarf (from cut pipes) sand etc. All these contaminants are eliminated or drastically reduced when the central heating system is installed by a competent installer. Furthermore correct flushing procedures at the time of installation (and subsequently) should eliminate this unnecessary corrosion factor *(see Flushing and Protecting – 2.5)*.

Corrosion Problems

Sludge collecting in pump

Corrosion causes the central heating system to decay and will result in frequent breakdowns. Further, it will reduce the efficiency and the working life of virtually all central heating components e.g. combi boiler, radiator, pump, valves etc. The damage continues unseen until it is too late. If there are visible signs of rust on any part of the system something is wrong. Rust marks on radiators or pipes means there is a leak. As the system corrodes sediment will build up. Radiators in particular will suffer from this deposition and have a cold area. Panel radiators are made of pressed mild steel and stressed steel is more pone to corrosion. Corrosion will occur at sharp corners, seams and if not vented of air, at the waterline close to the top of the radiator.

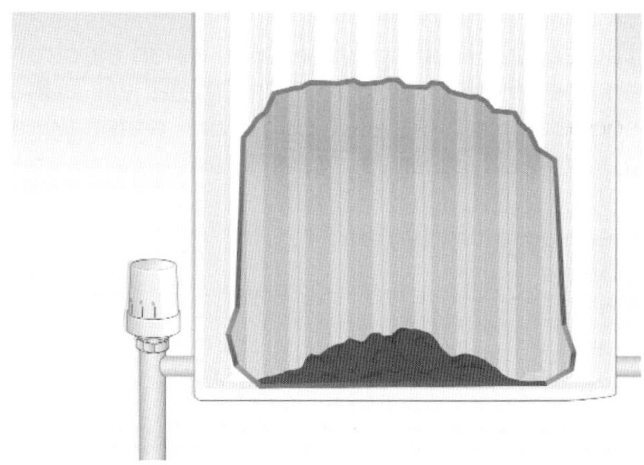

Sludge collecting in bottom of radiator

Test for Corrosion

Drain a small amount of water into a small container from the needle vent in the radiator. If it is reddish in colour all steel components may well be corroding from the inside out.

1. Fill a clean jam jar $^3/_4$ full with water from a radiator (drain from a vent plug which will be on the side at the top or back of the radiator) leaving $^1/_4$ air space in the jar.
2. Add 3 steel nails (not galvanized) to simulate the steel radiators and a penny/off cut of copper pipe to represent the pipe work.
3. Replace the lid tightly and seal with sticky tape and label it 'Jar A' and leave it for 1 week.
4. Repeat procedures 1 to 3 using a similar jam jar with water from the kitchen cold tap and label this 'Jar B.' This will act as a control for any corrosion and indicate if the problem to be remedied is with the cold water supply or within the central heating system.
5. At the end of 1 week check each jar. If there is rust on the nails in 'Jar A' then all the steel in the system is corroding. If the nails stay bright in 'Jar A' then the system is not suffering from corrosion. If the nails in 'Jar B' also exhibit rust the mains cold water supply to the property is either contributing to system problems or is the sole cause.

This corrosion process becomes more aggressive when the central heating system water is heated. The metal is being corroded away and collects at the bottom of the radiators, in the pipe work, heat exchangers and the lowest pipe work eventually blocking the flow of water. Lowest levels radiators and pipe work tend to be most affected. This sediment must be removed by correcting the fault, flushing the system and by using a good corrosion inhibitor chemical *(see Flushing and Protecting – 2.5)*.

Lime Scale Deposition
As fresh water percolates through the soil and underlying rock it slowly dissolves minerals, especially calcium and magnesium. The amount of dissolved minerals will depend on the geographical area the water is taken from. Water with a high mineral (hardness salts) content is commonly called hard water (conversely – soft water). When water is heated the dissolved bicarbonates break down to form carbonates and the higher the temperature to which this water is heated, the more carbonates will be formed.

$Ca(HCO_3)_2$	+ Heat	->	$CaCO_3$	+ CO_2	+ H_2O
Calcium Bicarbonate	+ Heat	->	Calcium Carbonate	+ Carbon Dioxide	+ Water
(Dissolved in Water)					

The insoluble carbonates are deposited within the central heating and domestic hot water system. The higher the hardness the more scale will be deposited. The higher the temperature to which the water is heated the more scale will be deposited. The higher the alkalinity of the water the greater the tendency for scale deposition. These deposits can be extremely detrimental to domestic central heating systems. The lime scale can cause blockages in valves, radiators, boilers, pumps etc. causing a reduction in efficiency of the system, increasing running costs and reducing the working life of the central heating components *(see Flushing and Protecting – 2.5)*.

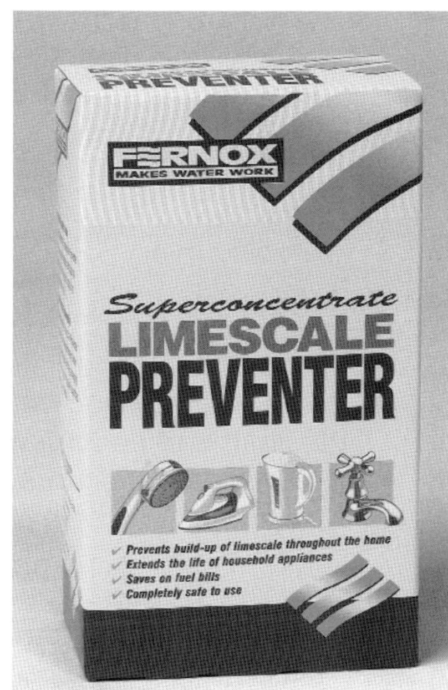

Should the hardness of the water exceed 200mg/litre it is good practice to treat the incoming mains water supply with a suitable base exchange water softener. This will protect the central heating system, all other water installations within the property and provide softened water. There is a pressure drop of about 1 bar with base-exchange water softeners so there must be sufficient mains water pressure (about 4 bar) minimum. Alternatively, an in-line scale reducer or ion exchange unit can be fitted on the cold water supply to the property or to the cold feed supply to the cylinder. Water softeners are often fitted without too many complications. Refer to manufacturer's instructions and recommendations for detailed fitting instructions.

2.5 Flushing & Protecting

Introduction

Corrosion within central heating systems can occur due to:-

1. Design faults
2. Electrolytic corrosion
3. Dezincification
4. Chlorine
5. Poor installation work procedures
6. Inadequate flushing at installation
7. Air entering the system

Many of the reasons for corrosion can be eliminated by good design and installation practices. Nevertheless, corrosion can still occur and it is wise to protect the central heating system with a chemical corrosion inhibitor.

To check for blockage in copper pipes (not in radiators or other ferrous fittings) touch a magnet to the pipe. If there is significant sludge in the pipe then the magnet will attract to the pipe. The stronger the attraction the more sludge build up there is within the pipe.

Evidence of system corrosion

Symptoms of Corrosion

1. Frequent venting of radiators because of hydrogen or air build up due to micro air leaks or 'pumping over'
2. Boiler noise which increases in frequency and intensity due to scale build up or black sludge (magnetite) deposition from rusting radiators
3. Pin hole leaks in radiators are an obvious sign that the radiators are rusting from the inside out
4. Pump or motorised valve breakdown(s) due to black sludge which, being abrasive, increases the shaft and bearing wear. The components either stick or fail completely
5. Cold spots in radiators due to a sediment of black sludge causing the lower central third of the radiator to become blocked and unable to freely pass heated water

Correcting the Cause

The cause of the corrosion must be diagnosed and eliminated before chemically treating the system. If it is aeration of a vented system eliminate 'pumping over' due to incorrect design installation *(see 'Central Heating – Fault Finding & Repair' book)*. If it is any of the above list numbers 2–6 flushing and protecting the system should greatly reduce corrosion problems. If air is entering the system this must be corrected *(see 'Central Heating – Fault Finding & Repair' book)*.

System Cleansing, Flushing and Protecting

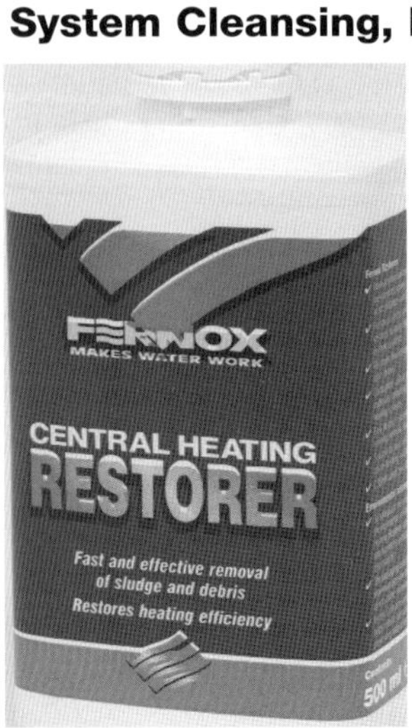

The objective when flushing and treating the central heating system is to remove scale, water suspended particles of sludge (magnetite) or rust (ferrous oxide) before chemically treating the water to inhibit corrosion or scale. Sludge and scale is very adherent and sticks to most internal surfaces of the central heating and hot water primary system. This needs to be loosened and removed before flushing. The best way to do this is by adding a system chemical to de-scale internal surfaces and break down sludge, moving it into suspension. When the central heating system water is drained the water suspended scale and sludge will also be removed.

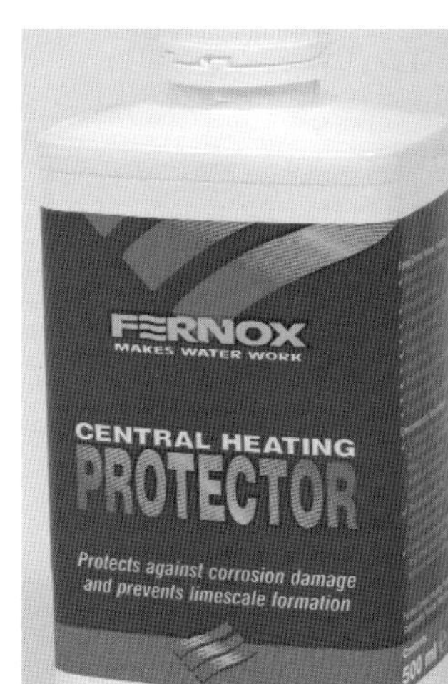

Chemical treatment of scale and sludge is not an immediate process so the chemical must be added to the system and left to circulate for up to 1 week depending on the severity of the problem. However, it is good practice to drain the existing system water first, refill and add the de-scaling/sludge remover chemical.

When all the debris, sludge and scale is removed, the system will be more efficient and contaminant produced noise from the boiler, pipes, radiators, pump etc. will be greatly reduced or eliminated.

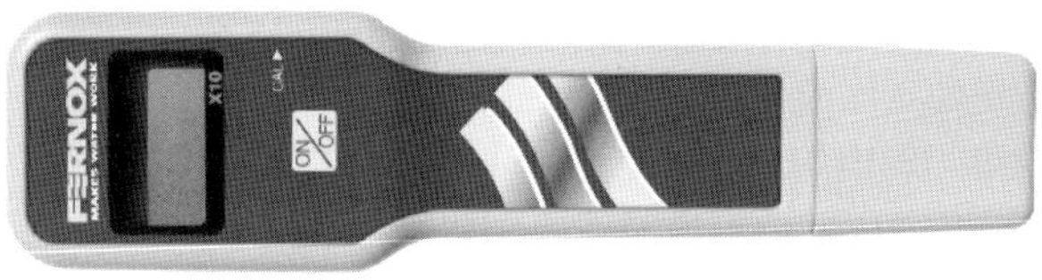

Total Dissolved Solids – TDS meter

The procedure below is also suitable for the pre-commission cleaning of new installations and combi boiler changes, removing flux, swarf, dust, sand etc. It is possible to test the system water with a chemical analysis kit supplied by the chemical manufacturer. However, these kits are only suitable if you know which chemicals may be in the system. If you do not know this then an indication of the level of corrosiveness of the system water can be gained by carrying out the nails and jam jar test *(see Corrosion and Lime Scale – 2.4)*.

1. Drain the entire central heating system including any drop feed radiators
2. Refill the system
3. Add the cleaner via a radiator air vent or into the feed and expansion cistern
4. Always circulate for at least 1 hour at normal operating temperatures
5. Allow system to operate in the normal heating cycle for about 1 week (this step is not necessary for the pre-commissioning of new installations)
6. Drain and flush the system at least twice or until the water runs clear
7. Refill the system adding a chemical corrosion inhibitor so the system is protected

Note:- Always follow the manufacturers instructions

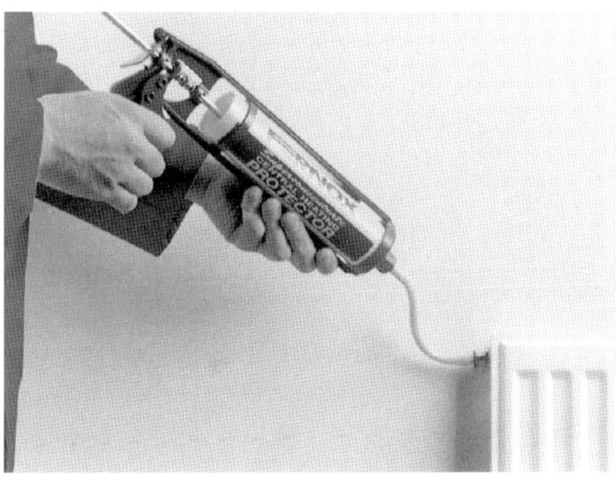

Injecting system chemicals via radiator vent

If the only concern is protecting the system from corrosion then just add the chemical. With an open vent system add it to the feed and expansion cistern, DO NOT accidentally add it to the cold water storage cistern. With a sealed system, as found in most combi boiler installations, inject the chemical via a radiator vent valve using the special attachment supplied with the chemical tube.

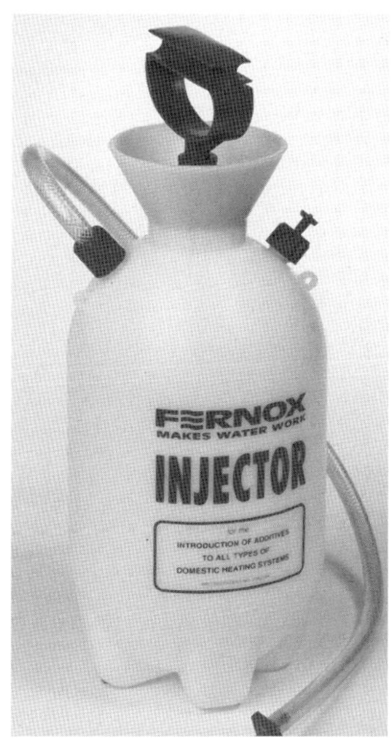

Chemical injector

Power Flushing The Combi Boiler and Heating System

Power flush machine

Power flushing is the term used when a central heating system, including the combi boiler, is cleaned using chemicals and a powerful self contained flushing unit. The principle is to create a powerful fresh water flow under controlled conditions to remove contaminants from the system. Quality power flushing units provide maximum flow at minimum pressure. This is the ONLY way to be certain the system is properly cleaned of sludge, scale and installation debris such as dust/flux/swarf and other contaminants. Power flushing is the most efficient and effective method of cleansing a central heating system. Pre-commission cleansing is a requirement for new systems and boiler replacements.

When a system has been power flushed the working life of the combi boiler and all its wet components such as the heat exchanger, pump, diverter valve, secondary heat exchanger etc along with the radiators and radiator valves is maximized. The better quality power flushing units efficiently remove sludge and scale and can:-

- Perform separate boiler descaling
- Eliminate boiler and circulation noise
- Clean the entire system in one visit
- Thoroughly clean the system by using a reversible flow action
- Provide a high flow with low pressure reducing the potential for uncovering system leaks

Installation Requirements

Contents

3.0 Installation Requirements

Introduction

When installing a combi boiler there are many regulations and requirements which must be complied with. This is probably summed up by the following statement:-

It is the law that a competent person in accordance with the gas safety (installation and use) regulations installs all gas appliances. Failure to install appliances correctly could lead to prosecution. It is in your interest and that of safety to ensure compliance with the law.

The above statement applies to combi boilers and the competent person referred to is a person who works with gas appliances or fittings and that person must be a member of a class of person approved by the HSE (Health & Safety Executive). The Council for Registered Gas Installers (CORGI) operates such a scheme *(see Gas Safety – 4.1)*.

All installations of combi boilers must be in accordance with the relevant requirements of the Gas Safety Regulations, IEE (Wiring) Regulations, Building Regulations, Water Regulations and of course the Manufacturers instructions.

Professional codes of practice set best practice standards for the correct installation, commissioning and servicing of combination boilers and associated system components *(see Flushing & Protecting – 2.5)*.

It is a regulation that when a combi boiler is connected to the gas supply it must be tested in accordance with the manufacturers instructions i.e. commissioned.

There are mandatory requirements that apply to combi boilers, flues, flue terminals, ventilation, water connections and wiring. The following sections describe in depth many of the requirements but are not exhaustive. It is not possible for one book to identify and describe all the regulations which must be observed when installing a combination boiler.

Introduction

Most installers, customers and specifiers prefer combi boilers with fan assisted balanced flues. Combi boilers flues and terminals are no different to the flues and terminals used by system boilers therefore all of the variations will be encountered. The variations of flues, flue runs and terminals is vast and will include:-

Open Flue (conventional flue)
With Diverter incorporated within the combi boiler
With Diverter fitted separated to the combi boiler
Balanced Compartment
Room Sealed
Natural Draught
Fan Assisted
Forced Draught
Vertex Flue
Two Pipe Flue
Shared Flue

Flue pipes are available in an overwhelming range of sizes and construction materials including vitreous enamelled mild steel, stainless steel, aluminium, fireproof materials and for many condensing combi boilers graded plastic *(see Condensate Discharge System – 6.4)*.

The flue run may be horizontal or vertical passing through ceilings, through floors, close to floor joists, internal and external walls and in some cases passing through a roof. The external walls may be un-insulated or insulated cavity type or of timber frame construction. The insulation may be glass or rock fibre batts, loose fill polystyrene granules or boards, formaldehyde (UF) foam etc. The walls of timber frame properties may have a plaster board, a vapour control layer, insulation, timber sheeting, breather paper and of course the external brick work.

Before undertaking any flue work, try and establish with the customer the wall construction and insulation type, if any. Fortunately for us 98% of all domestic combi boilers are wall mounted room sealed fan assisted balanced flue installations. Again, fortunately most of these will be mounted internally on an external wall.

Room sealed combi boilers take the combustion air from outside the property and in most cases this is where the products of combustion are discharged. As the flue terminal is taking in air and discharging flue gases at a point where the atmospheric pressure is the same, the term 'balanced flue' is used. Room sealed balanced flue combi boilers should be the combi boiler installation of choice.

Flue Systems

Typically, off the top of the combi boiler will be the flue turret connected to a 100mm diameter telescopic flue duct assembly which passes through the wall. The flue assembly has a typical maximum length in the region of 700mm but may be extended if the install is a side flue opening. Where the flue is to pass through a cavity wall a hole is either stitch drilled or a 107mm core drill is used.

Fanned-flue combi boilers can be connected to either a horizontal or vertical flue system. These two systems can be either:-

Co-axial or concentric pipe systems
Parallel or twin pipe systems

Co-axial Flue System

Using a 100mm diameter concentric flue and air pipe they usually comprise a 60mm pipe located centrally with a 100mm pipe. The 60mm inner pipe channels the flue gases from the combi boiler to outside the property and the 100mm outside pipe brings fresh air from the atmosphere to the combi boiler.

Co-axial systems use clamp or push fit connections, occasionally a mixture of both and can have a flue run up to 4 metres. Usually supplied by the combi boiler manufacturer the maximum run varies with type of combi boiler, the number of bends to be installed and is longer if 125mm diameter flue is used.

Condensing Combi Boilers – The flue system is similar to a non condensing combi boiler (conventional) system but there are important differences in both the material used and often the installation method. To allow the condensate to drain back to the collector trap within the combi boiler the flue should be installed with a continuous fall back to the combi boiler. For short flue runs some manufacturers produce a co-axial flue system with this fall already built in to the inner 60mm pipe.

Conventional Combi Boiler – The flue system should have a continual fall to the outside to reduce the possibility of rain water entering the flue and combi boiler.

Parallel Flue System

Rarely seen in the domestic market, parallel flue systems are increasing in popularity with the advent of the high efficiency condensing combi boiler only regulation. Allowing longer flue runs not all combi boilers are suitable for parallel systems.

Comprising two individual pipes connected either directly or indirectly to the boiler using a connection kit. The connections are push fit sealing type which ensures there is no leakage of flue gases or condensate. At the point of termination a converter box is normally used which converts the parallel flue pipes to a concentric system. When this system is used with non-condensing combi boilers it is ALWAYS necessary to install a condensate collector/trap.

Flue Terminals

If the terminal of a balanced flue room sealed combi boiler is sited too close to an obstruction, the products of combustion may not disperse. The incoming combustion air may be contaminated by the flue gases (vitiation) resulting in a lack of oxygen to the main burner and incomplete combustion. The flames may 'lift off' in search of oxygen and may further cause the combi boiler to shut down or go to lock out. This nuisance shut down causes customer dissatisfaction. Avoid if possible siting a terminal in a car port, under a balcony, in a light well, near security lighting or where it might be blocked by snow.

The terminal location must satisfy all the appropriate regulations. By complying with the regulations all the products of combustion will be totally cleared and will also prevent:-

Inefficient combustion of fuel
Flue gases entering the dwelling and/or neighbouring properties
Fire hazards
Staining to walls by the flue gases
Upsetting the customer!

Note:- If a balanced fanned flue terminal is sited less than 75mm below a plastic gutter or 200mm below painted or UPVC eaves a 1 metre long heat shield must be fitted. Remember to leave an air gap of at least 5mm between the surface and heat shield.

Note:- A flue terminal guard must be fitted over the terminal if the terminal is less than 2 metres from a finished floor level. No part of the guard should be less than 50mm from any part of the terminal, any openings should be no larger than 16mm and have no sharp edges. The guard should be fixed with plated screws.

Note well:- Terminal position measurements change with regulation changes and as such ALWAYS refer to the appropriate regulation.

Boundaries

Flue terminals must be positioned so the products of combustion do not cause a nuisance to the inhabitants of adjoining or adjacent properties. In cold weather the flue gases may condense in the cold air forming a cloud of water vapour. This is known as 'pluming' and may be a further cause of nuisance to neighbours.

Should the neighbour formalise their reasonable complaint and it goes to court you will probably lose the argument. You will then be ordered to re-site the terminal, which probably means moving the combi boiler, all at your own expense! ALWAYS carefully consider flue terminal locations.

Service

At service check the flue to ensure that it is complete, is the correct size, has no obstruction and is good sound condition. Check the terminal is the correct size, is correctly sited, is of an approved type and fitted with appropriate guard if necessary.

Remember:- Always follow the Manufacturers Instructions

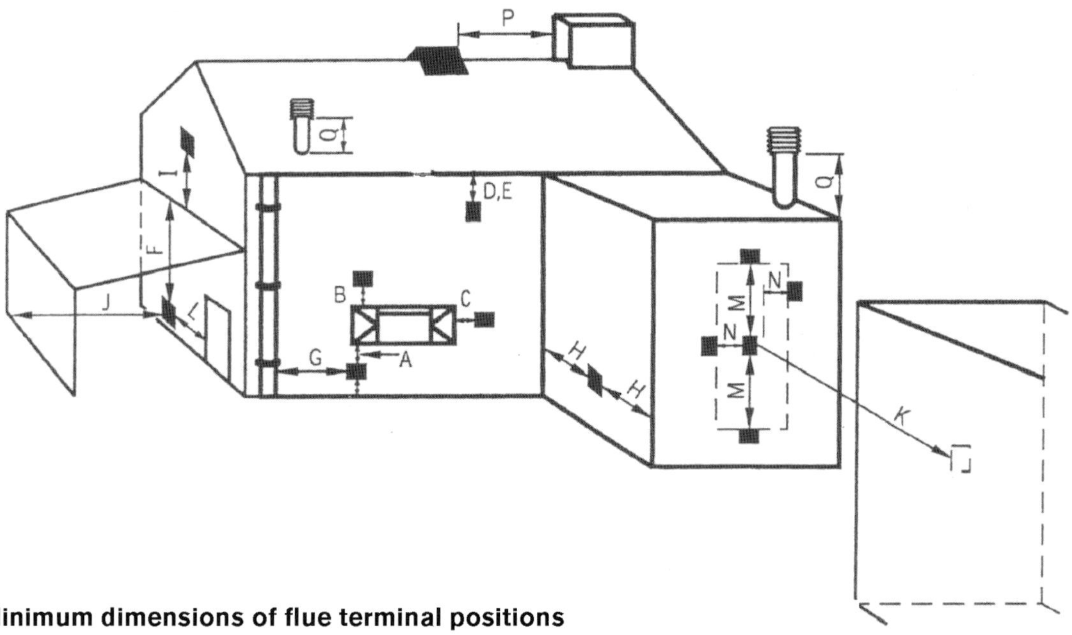

Minimum dimensions of flue terminal positions

Dimension	Terminal Position (kW input expressed in net)	Balanced flues room sealed: Fanned draught
A 1)	Directly below an opening, air brick, opening windows, etc.	300 mm
B 1)	Above an opening, air brick, opening window, etc.	300 mm
C 1)	Horizontally to an opening, air brick, opening window, etc.	300 mm
D	Below gutters, soil pipes or drain pipes	75 mm
E	Below eaves	200 mm
F	Below balconies or car port roof	200 mm
G	From a vertical drain pipe or soil pipe	150 mm
H	From an internal or external corner	300 mm
I	Above ground roof or balcony level	300 mm
J	From a surface facing the terminal	600 mm
K	From a terminal facing the terminal	1200 mm
L	From an opening in the car port (e. g. door, window) into the dwelling	1200 mm
M	Vertically from a terminal on the same wall	1500mm
N	Horizontally from a terminal on the same wall	300 mm
O	From the wall on which the terminal is mounted	Not applicable
P	From a vertical structure on the roof	Not applicable
Q	Above intersection with roof	Not applicable

1) In addition, the terminal should not be nearer than 150 mm (fanned draught) to an opening in the building fabric formed for the purpose of accommodating a built-in element such as a window frame.

Introduction

Nearly all combination boilers are room sealed and therefore do not require any permanent unrestricted air flow via a purpose provided air vent(s).

Whenever and wherever possible it is always preferable to fit a room sealed combi boiler using a balanced flue, keeping the flue run as short as possible and the flue terminal mindfully sited *(see Flues & Terminals – 3.1)*.

Open Flue Ventilation

There are open flued combi boilers installed and these must have purpose provided ventilation. This ventilation will provide clean fresh air ensuring complete combustion, assist the draw performance of the open flue, remove condensation and provided cooling air. Some of this air will be provided through naturally occurring sources e.g. gaps around doors and windows, floors etc and an allowance is made for this. This natural ventilation air is known as 'adventitious air' and is deemed to provide ventilation air for the first 7 kW (gross or net) of a combi boiler heat input rating. The room in which the combi boiler is installed must have an air vent which provides an effective free area of $5cm^2$ (net) or $4.5cm^2$ (gross) for every kilowatt rating in excess of 7 kW.

The data badge on older open flue combi boilers may give the boiler rating in kW gross so to convert to kW net divide by 1.1.

Example:- Open flued combi boiler rated at 29.5 kW gross 29.5 divided 1.1 = 26.8 kW net.

A 27 kW (net) open flue combi boiler would therefore require a vent of:-

27 kW (net) – 7 kW (adventitious air) = 20 kW   20 kW x $5cm^2$ = $100cm^2$ ($10{,}000mm^2$)

Vent manufacturers make a wide range of air vents but it is not limitless so the ventilation requirements will have to be rounded up to the nearest vent size available. Ventilation through cavity walls or through timber framed walls MUST incorporate a continuous sleeve from the inside to the outside. It is sensible to slightly incline the sleeve to prevent water ingress.

Ensure the vents are free of blockages such as dust, debris, cobwebs etc, remove fly screen if fitted, ensure the householder has not blocked off the vent with sticky tape/cling film and so on. If the vent is of a closeable type either replace with an appropriate air vent or fix permanently in the open position. The vent must be positioned where it cannot be blocked by snow or vegetation.

Finally, if the ventilation is unsatisfactory the appliance must be categorised as 'at risk' and should not be used until the ventilation faults are corrected.

Cooling

Room sealed combi boilers do not require ventilation air for combustion purposes but when installed in a cupboard may require ventilation air for cooling purposes. The manufacturers instructions specify minimum clearances around and in front of the combi boiler and MUST be followed. These clearances are also specified to allow service and repair to the appliance.

Most modern room sealed condensing combi boilers have a maximum surface temperature of casing and the flue of less than 85°C. This means that no specific safety precautions are required with regard to flammable building materials and fitted furniture like kitchen cabinets. However, the specified clearances must be maintained. When installed in a cupboard/compartment the low casing losses from the appliance eliminate the need for purpose provided ventilation openings.

When an appliance is installed in a cupboard/compartment which is to be used for airing clothes the airing space must be separated from the boiler space by a perforated non-combustible partition. Expanded metal or rigid wire mesh is acceptable provided the major dimension is less than 13mm.

Typically, modern room sealed, balanced flue condensing combi boilers require clearances at the side of the appliance as low as 5mm, 30mm clearance above the top of the flue turret and 200mm clearance under the appliance. To the front of the combi boiler 75mm with the cupboard door closed and a minimum of 600mm with the door open which allows access for service and repair.

Remember:- Always follow the Manufacturers Instructions

3.3 Water Regulations

Introduction
The water regulations make provision for:-

'Preventing contamination, waste, misuse, undue consumption and erroneous measurement of water supplied by a water undertaker'

Installers and engineers therefore have a duty to comply with the water regulations where the regulations relate to any work they may carry out whilst installing, servicing or repairing combi boilers.

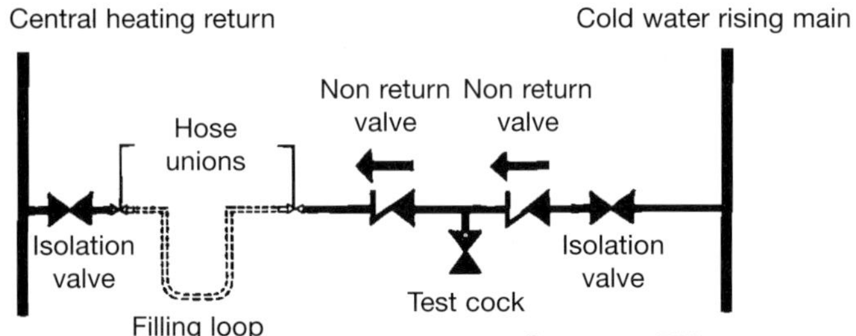

System Fill

The regulations specifically apply to methods used for filling and 'topping up' the central heating system such as 'filling loops' *(see Filling Loops – 6.8)*. Also, the regulations refer to the waste of water and heat energy lost with long hot water pipe runs i.e. 'active' dead legs *(see Dead Legs – 2.3)*.

Back Flow
As all the cold and hot water draw off points are effectively connected directly to the cold water mains supply, the installer must ensure compliance with regulations at all points.

The head of a loose head shower must not be allowed to fall closer than 25mm above the top edge of the bath. This will prevent the shower head being immersed in bath water where back flow is then a risk. Alternatively, the shower must be fitted with a retaining ring or an anti siphonage device at the point of flexible hose connections. DO NOT install a non return valve on the supply to an electric instantaneous shower. This can seriously damage the shower unit.

The supply of hot and cold mains water direct to a bidet is acceptable provided the bidet is the over-rim flushing type. The outlet(s) should be shrouded and unable to have any hand held spray assembly attached. No anti-siphonage arrangements are required.

Energy Conservation
When a hot water draw off point is located some distance from the combi boiler a long pipe run is necessary. This pipe run is known as an active dead leg *(see Dead Legs – 2.3)* as hot water flows from the combi boiler to the draw off point, heat energy is lost.

It is a regulation that all water fittings forming part of a secondary hot water system and all pipe work carrying hot water to a draw off point that are longer than the maximum length shown below should be thermally insulated:-

Maximum recommended length of un-insulated hot water pipe work

Outside diameter of pipe (mm)	maximum pipe run (Metres)
Over 12 to 22 (inc)	12

Therefore, should a 15mm pipe run be longer than 12 metres (including fittings) ALWAYS insulate as much of the pipe as is practicable. Insulate pipe work as it passes through or under:-

Kitchen cabinets	Cupboards	Under stairs cupboard	Floorboards and voids
Roof spaces	Stud walls	Baths and shower trays	

Introduction
When carrying out a site visit for a proposed combi boiler installation it is ALWAYS necessary to check the electrics of the property. If there is any doubt of the integrity of the electrical installation, use the service of a NICEIC Registered or competent electrician.

Using a quality socket tester or ring tester will show if the property has reversed polarity or other fault. If it has, again use the services of a NICEIC Registered or competent electrician to correct the fault. The socket tester will check but not measure the following:-

Earth path is within limits

Voltage is correct

Polarity is correct

It is also prudent to use a socket tester when called to a break down or before a combi boiler service. If there is an electrical fault DO NOT attempt to repair or service the combi boiler. Electrically isolate the combi boiler, inform the householder, in writing, of the possible wiring fault and inform them to use the services of a NICEIC Registered or competent electrician.

Electrical Bonding
All domestic gas installations MUST have, by law, main equipotential bonding of the gas pipe work installed by a competent person. This is also called cross bonding or permanent cross bonding. By connecting all metal pipe work a zone is created in which the voltage difference and therefore the risk of electric shock is minimised.

This simply means that electrical cross bonding will be used. The bond must be fitted by a clamp to the gas pipe not more than 600mm from the meter outlet i.e. on the customer's side of the meter and fitted inside the building. The clamp must have a warning tag or label stating 'Safety Electrical Connection. Do not remove.' The bond will have a minimum cross sectional area of at least $10mm^2$. Check it runs to the consumer unit and finally check the consumer unit is bonded to the mains supply with a bond of at least $16mm^2$ C.S.A. If in any doubt use the services of a NICEIC Registered or competent electrician. Further, the bond should make a mechanically and electrically sound connection and must also be resistant to corrosion.

Supply Requirements
Combi boilers must have a means of isolation from the electrical supply. There are three ways of achieving this:-

Switched Fused Connection Unit – The combi boiler can be connected directly to a switched fused connection unit fitted with a 3 Amp Fuse. This may be wired into an existing socket ring main using flat twin core and earth cable with a minimum cross sectional area of $2.5mm^2$. The earth in this cable must have a minimum cross sectional area of $1.5mm^2$. This is a standard cable and is readily available. Where the cable is surface run it must be supported with cable clips. When buried in wall ALWAYS run the cable through trunking either vertically or horizontally. Be aware that there may be buried diagonal cable runs.

The switched fuse connection unit may also be wired directly into the consumer unit using flat twin core and earth cable with a minimum cross sectional area of $2.5mm^2$. This is known as a radial circuit. The earth in this cable must have a minimum cross sectional area of $1.5mm^2$. This is a standard cable and is readily available. Where the cable is surface run it must be supported with cable clips. When buried in walls ALWAYS ensure the cable is run through trunking either vertically or horizontally. Be aware that there may be buried diagonal cable runs. The radial circuit must be protected at the consumer unit with a 5 Amp fuse or miniature circuit breaker (MCB). The wiring installation procedure must be carried out by a NICEIC Registered or competent electrician.

3.4 Electrical/Wiring

Plug and Socket – The combi boiler can be electrically supplied via a standard switched or unswitched single or double power outlet socket. The standard 3 pin plug must be rated to 13 Amp minimum and fitted with a 3 Amp fuse. Electrical standards recommend the use of a switched fused spur i.e. switched, fused connection unit.

The combi boiler is connected to the fused connection unit or plug using round 3 core flexible cable with a minimum cross sectional area of 1mm². Keep the cable run as short as possible, use cable clips to support the cable and to give a professional look to the installation. Heat loss through the casing of modern combi boilers is extremely low, however, the central heating pipes do get hot, up to 82°C so using heat resistant cable is good practice. Avoid running the cable next to hot pipe work.

Room Thermostat/Programmer

All combi boilers should have a room thermostat or programmable thermostat installed to provide 'boiler interlock' preventing 'boiler cycling.' By preventing boiler cycling the working life of the combi boiler and its components will be maximised. It is also an effective way to achieve maximum efficiency and minimise running costs.

The installation of a room thermostat or programmable thermostat is a relatively straightforward procedure. The wiring is normally achieved by using flat twin core and earth cable with a minimum cross sectional area of 1mm². Some thermostat manufacturers produce a wireless version which eliminates all hard wiring to the thermostat. However, the receiver is normally sited close to the boiler and this must still be wired in.

When installing a room thermostat or programme thermostat always follow the combi boiler and control manufacturer's instructions. If you are not competent in wiring techniques always use the services of an NICEIC Registered or competent electrician.

Safety

Contents

4.0 Safety

Introduction

Some people consider safe working practice as avoiding bruising a thumb or cutting a finger. It is this but it is also much more. If you are carrying out a job/project/mission and during, or afterwards, you or someone else is injured or adversely affected (e.g. from drinking contaminated water) you have to prove that you, so far as is practicable, took all reasonable steps to prevent injury or damage to person or property. You would be unlikely to prosecute yourself in the case of injury but if someone else is injured and/or property damaged they, or their insurance company, may prosecute.

Avoid prosecution, avoid hurting yourself, avoid hurting others and avoid damaging property. Always follow safe working practices.

Professionals

As a professional you will know of a 'horror story' or two when it all went wrong. People die every year of injuries/accidents that could, and should, have been avoided. A safe working environment is essential at every job and it is your responsibility to get it right every time, avoiding injury to yourself and others and safeguarding property.

Introduction

The law stipulates that any person carrying out any work in relation to gas fitting must be competent to do so. Simply put, if you are not trained in gas fitting and do not know the regulations you must not carry out any work on gas pipes, gas fittings or gas appliances.

Any work with gas appliances or fittings must be carried out by a member of a class of persons approved by the HSE (Health and Safety Executive). The council for registered gas installers (CORGI) operates a scheme.

Gas Escape

If you smell gas:-

1. Turn off the gas supply immediately at the meter or emergency control valve
2. Extinguish all sources of ignition
3. Do not smoke
4. Do not operate electrical light or power switches or bell pushes (on or off)
5. Ventilate the building(s) by opening doors and windows
6. Ensure access to the premises can be made
7. Report the escape to the gas emergency service call centre on 0800 111 999 (N. Ireland – 0800 002 001)

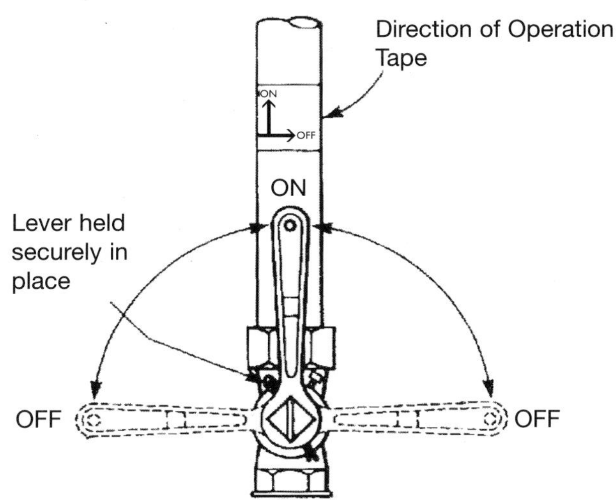

Emergency/Meter Control Valve

Notice to Practical Householders

Do not attempt gas work if you are not competent.

You must not use or allow to be used any equipment/appliance you think may be faulty. Have your gas appliances serviced regularly (yearly) by a competent person (CORGI registered). You must comply with the law.

It is a regulation that when an appliance has been connected to a gas supply it must be tested in accordance with the manufacturer's instructions. In other words,

'Commissioned'

Whether the manufacturer's instructions explicitly detail let by and tightness test and 'purging' of the gas pipe work these tests must be carried out when any work is carried out on the gas pipe work. The let by and tightness test *(see Let By Test – 5.6 and Tightness Test – 5.7)* can be carried out in a few minutes with purging of the pipe work *(see Tightness Test – 5.7)* again only taking a few minutes. The appropriate safety procedures must be adhered to.

If there is a smell of gas, a gas leak detector *(see Flue Gas Analysers – 5.3)* can be used to trace the escape.

Carbon Monoxide

Carbon monoxide (CO) is a colourless, odourless and tasteless gas which is a product of incomplete combustion of carbon based fuels. If breathed in it is harmful and if absorbed into the body in sufficient quantities (less than you think!) can result in death.

What makes carbon monoxide poisoning dangerous is it goes unnoticed. It affects reasoning, mental ability and coordination which contributes to preventing escape from the carbon monoxide source. Exercise increases its effect on the body. The major sources of carbon monoxide are motor vehicles exhaust fumes and the burning of carbon and hydrocarbon fuels e.g. wood, coal, smokeless fuels, oil, natural gas, LPG etc. All fires, room heaters, boilers, cookers, portable heating, fuelled by natural gas, LPG, oil (paraffin) need an adequate supply of fresh air to ensure complete combustion and therefore minimal formation of carbon monoxide.

If the appliance is poorly maintained or there is poor ventilation problems can occur. If the appliance or flue/chimney is faulty, again high and potentially lethal levels of carbon monoxide can be quickly produced creating an 'Immediately Dangerous' situation.

The danger signs on appliances include:-

1. Staining of the appliance or room decoration
2. Excessive yellow or orange flames
3. Soot deposits
4. Unstable combustion

Do not block air vents. Keep them clean and clear of cobwebs etc.

With wood, coal and smokeless fuel it is not possible to achieve complete combustion as air (oxygen) is not mixed in the same way as with liquid and gas fuels. Therefore, when burnt, these fuels will always produce very high and dangerous levels of carbon monoxide. Ordinarily this dangerous product of incomplete combustion is safely removed via the flue/chimney but if it is blocked extremely high levels are produced which can cause death in minutes *(see table below)*.

Effects of carbon monoxide

Parts per million	Effects
50	Exposure limit is 8 hours (average healthy person)
200	Headache after 6 hours at rest or 2 hours exertion
400	Headache and discomfort and possible collapse after 2 hours at rest or 45 mins exertion
1200	Palpitations after 30 mins at rest or 10 mins exertion
2000	Unconscious after 30 mins at rest or 10 mins exertion

Parts per million:- Low concentration of gases when measured in air are measured as parts per million i.e. 1 molecule (unit) of the gas per million molecules (units) of air.

Alternatively, to show how very small quantities of carbon monoxide produce extremely adverse effects on people is by expressing the % volume of carbon monoxide in the air *(see table below)*.

% volume of CO in air	Symptoms
0.01%	Tightness across the forehead with exposure of 2 to 3 hours
0.03%	Flushed skin, headache, breathlessness and palpitation on exertion within 4 to 5 hours exposure
0.05%	Severe headache, nausea, weakness of limbs, irritability, increased respiration and pulse rate followed by collapse on exertion within 4 to 5 hours exposure
0.4%	Coma, weakened respiration and heart, possible death within 10 mins exposure

In the normal breathing process oxygen combines with haemoglobin in the red blood cells which transport it around the body as oxyhaemoglobin to where it is needed. The oxygen is given up by the red blood cells to the tissues and the process starts again. When carbon monoxide is breathed in it reacts with the haemoglobin in a similar way but forms carboxyhaemoglobin. The bond between carbon monoxide and haemoglobin is far stronger than between oxygen and haemoglobin so it does not break away easily and stays attached for longer. Carboxyhaemoglobin cannot carry oxygen so if sufficient is formed the body will be starved of oxygen, asphyxia will result and the exposed person will die.

Fortunately the bond between carbon monoxide and haemoglobin is not permanent so sudden high doses are completely reversible and a full recovery will be made. When fresh uncontaminated air is breathed in the red blood cells will eventually give up the carbon monoxide and be replaced with oxygen. Normal oxygen levels in the body are then gradually restored. However, long term exposure to low levels of carbon monoxide will produce high levels of carboxyhaemoglobin in the blood. These sustained levels can attack the proteins in the body and cause irreversible damage to vital organs including the brain *(see table below)*.

% saturation of haemoglobin with carbon monoxide	**Symptoms**
0.1–2.6%	Natural level
5–15%	Cigarette smoker
10–20%	Tightness across forehead
20–30%	Flushed skin, headache, breathlessness, palpitations
30–40%	Severe headache, dizziness, nausea, weakness of knees, irritability, impaired judgement
40–50%	As above with increased perspiration and pulse rates, collapse on exertion
50–60%	Loss of consciousness, coma
60–70%	Coma, weakness respiration and heart
Over 70%	Respiratory failure and death

Carbon monoxide poisoning and common flu symptoms are similar at first. Early warning signs of low-level carbon monoxide poisoning include tiredness, headaches, dizziness, nausea, vomiting and shortness of breath. In about half of carbon monoxide poisoning cases the skin turns pink or reddish in response to an increase in blood pressure.

Should a person or you describe the symptoms above and you suspect carbon monoxide poisoning you must:-

a) Vacate the property
b) Immediately seek professional medical attention
c) Immediately discontinue use of any gas, oil, and/or solid fuel appliances
d) Contact competent gas/oil/solid fuel installers or engineers to check the installation

Remember that young, old and sick people are more at risk even at lower levels.

BEWARE CARBON MONOXIDE. FIT A CARBON MONOXIDE DETECTER OR ALARM

Electric Shock

The effects of electric shock vary from person to person. A current of over 75mA (milli ampere – 75/1000th of an ampere) might be fatal. A sensation of shock is felt at just 1mA (1/1000th of an ampere). The higher the current (milli amps) and the longer the contact the greater the risk. So, a current of say 300mA (300/1000th of an amp) for 30 milli seconds (30/1000th of a second) should have no lasting ill effects. However, a current of say 50mA (50/1000th of an amp) for one second may be fatal in certain circumstances.

These circumstances may dictate the amount of current able to pass through the body. The higher the resistance of the body the lower the shock current. So the resistance given by gloves, footwear, flooring (say carpet) and the conditions (dry hands, dry floor, dry central heating components) will increase the total resistance therefore reducing the shock current. Conversely a low total body resistance i.e. no gloves, wet hands, thin or no footwear combined with a wet floor and wet components will reduce the body resistance to current and greatly increase the risk of serious or fatal electric shock.

With this in mind always ensure the appliance is correctly fused at the plug, all cables are undamaged and securely fixed. Cables should also be kept away from heat sources e.g. hot pipes, lamps etc.

Never work on a live circuit. Always turn off the supply before commencing any work that may involve electricity. Test the circuit is off by using a multimeter *(see Multimeters – 5.4)*. If necessary remove the fuse from the connection unit plug and/or consumer unit and keep it on you until you are ready to replace it.

Electrically Isolating The Combi Boiler

Method One – If a 3 core flexible lead is wired to a protected terminal block on the appliance and the other end is wired to a 3 pin plug inserted into a socket then :-

a) Switch off at the socket (if it has a switch)
b) Remove plug from socket
c) Remove the fuse from the plug (and put it in your pocket for safe keeping)
d) Tie up the flexible lead so it cannot reach the socket

Method Two – If the internal wiring of the appliance is terminated by a 'Bulgin' type plug connector socket which is fixed to the housing of the appliance and the other end wired to a 3 pin plug inserted into a socket then:-

a) Switch off at socket (if it has a switch)
b) Remove plug from socket
c) Withdraw Bulgin plug connector
d) Place plug connector and flex in tool box

Method Three – If the appliance is permanently wired to the house circuit it should have a double pole isolating switch with a fuse so:-

a) Switch off isolation switch (if it has a switch)
b) Remove fuse (if it has a fuse)
c) Put the fuse in your pocket (for safe keeping)
d) If there is no fuse then tape over the switch in the off position and put a notice on it informing people the circuit is being worked on i.e. 'Do Not Switch On'
e) If there is an ordinary single pole switch, switch the circuit off at the consumer unit and remove the fuse
f) DO NOT put the fuse on top of the consumer unit, put it in your pocket for safe keeping
g) Return to the combi boiler and check across the terminals for zero voltage

Electrical Safety Tests

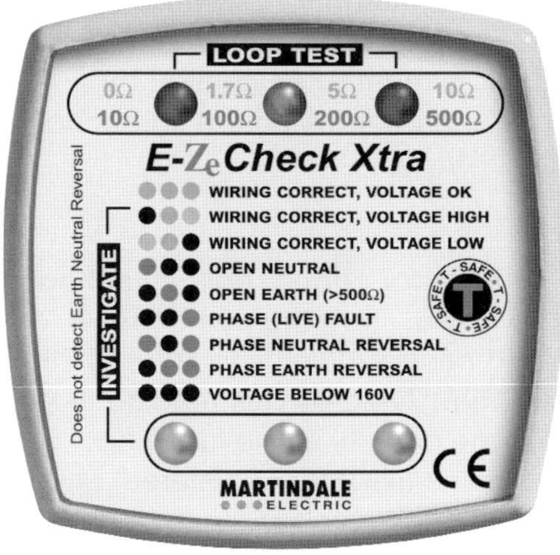

Socket tester

To ensure safety for you and the customer there are 5 essential electrical safety tests that must be carried out before starting any combi boiler fault finding and repair. These 5 tests MUST be carried out EVERYTIME a combi boiler is worked on. Even if you installed the combi boiler ALWAYS carry out these tests as the property may have been re-wired or partially re-wired, the combi boiler may have been worked on etc. If the works were carried out by a competent person then your 5 tests will prove the electrical safety of the combi boiler. The works may have been carried out by an untrained person or possibly the householder and if you do not carry out these tests you may discover their incompetence, the hard way.

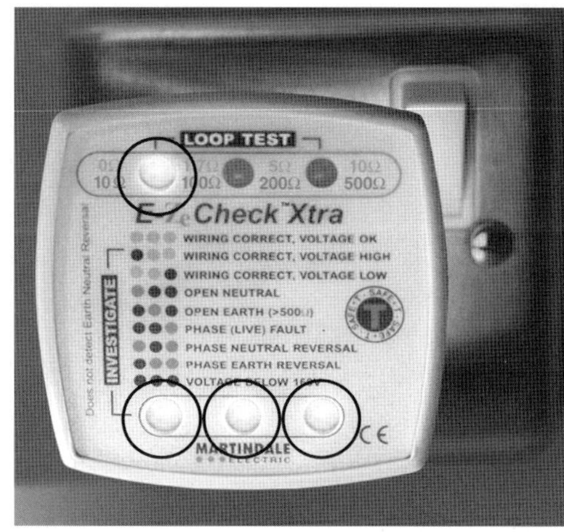

Socket tester with proving neons

Before carrying out electrical tests you must be competent to do so. The five tests are:-

1. Voltage Test
2. Earth Continuity Test
3. Short Circuit Test
4. Resistance To Earth Test
5. Polarity Test

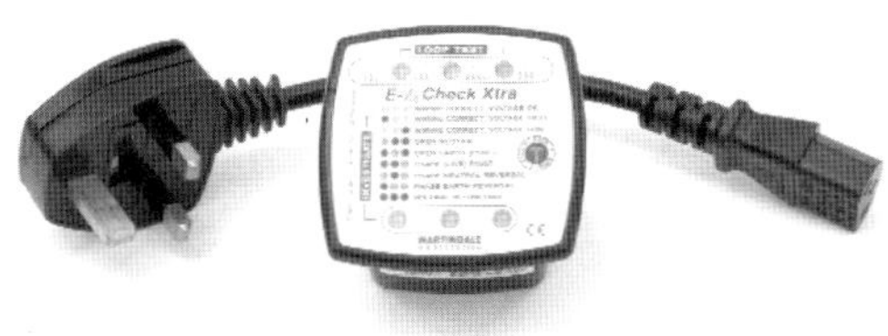

Socket tester with adapter

Multimeter

Voltage Test

The combi boiler must be electrically isolated before work is commenced. To confirm there is no voltage to the combi boiler test at the terminal block between live and neutral. There should be zero volts *(see Multimeters – 5.4).*

Earth Continuity Test

The combi boiler must be electrically isolated. This test confirms earth continuity between the earth supply to the combi boiler and the earth connections and casing of the combi boiler. The test is carried out between the appliance case (also test the pump, fan, etc) and the retaining screws of the socket or fused connection unit. The resistance MUST be no more than 1 Ohm (Ω). If the resistance is greater than 1 Ohm the fault must be rectified before carrying out any further tests as any metal parts that should be earthed may become live and in the event of a short circuit cause an electric shock *(see Multimeters – 5.4).*

Short Circuit Test

The combi boiler must be electrically isolated. This test confirms the integrity of the internal resistance of the combi boiler. The test is carried out between the live and neutral connections and a reading of less than 100 Ohms (Ω), with a 3 Amp fuse fitted, indicates a short circuit condition. The desired reading is Mega Ohms (MΩ), Overload (O/L) or Infinity (∞). The fault condition is often marked by repeated fuse blowing and/or visual evidence of localized burning/scorching. It is most often caused by wires being trapped or pinched destroying the insulation. With no visual evidence individual components should be isolated from the electrical circuit and the test repeated *(see Multimeters – 5.4).*

Resistance To Earth Test
The combi boiler must be electrically isolated. This test confirms the integrity (resistance) of the insulation between live and earth. The test is carried out between live and earth connections and a reading less than 1 Mega Ohm (MΩ) indicates a fault. Again, this fault is most often caused by a wire being trapped/pinched destroying the insulation. Note:- All switches and controls on the combi boiler must be in the 'on' position, as if calling for heat *(See Multimeters – 5.4)*.

Polarity Test
This is the last of the 5 tests to be carried out and it is a LIVE test. It tests for correct polarity although some combi boilers will still function satisfactorily when polarity is reversed. Turn the combi boiler 'on' but turn all controls 'off' so there is no demand. Test between live and neutral which should read 230V (approx). Test between neutral and earth which MUST be less than 15V. Typically this reading is less than 4V, however, a reading greater that 15V indicates a fault. Check the polarity at the fused spur and if there is still a fault the problem is within the house wiring. An electrician (NICEIC Registered) should be called in to rectify the house wiring fault *(see Multimeters – 5.4)*.

Temporary Earth Continuity
When working on gas pipes or pipes that carried gas there is a risk of:-

1. Ignition or gas escape
2. Electric Shock

Therefore, to prevent any possibility of fire, explosion or gas escaping when work is carried out on gas pipes the meter MUST be temporarily disconnected, pipe work sealed and dust caps fitted to the meter.

Further, it is possible for gas pipes to act as an earth return path when electrical faults are present. So earth continuity MUST be maintained at all times. This can be achieved by the use of a temporary continuity bond usually supplied with crocodile clips attached and insulated handles. This should be at least 1.2 metres long of single core multi strand, insulated flexible cable with a minimum cross sectional area of $10mm^2$. The temporary bond MUST be used when the meter is temporarily disconnected or breaking or cutting pipe work.

The following is the correct process for using a temporary continuity bond:-

It is advisable to isolate any gas appliances from the mains electrical supply
Turn off the gas supply
Establish where the continuity bond will be attached and remove paint/coatings/dust so a good connection is possible

Connect the continuity bond:-
First – upstream side of meter (service pipe side) or pipe work to be worked on
Second – downstream side of meter (appliance side) or pipe work to be worked on

Carry out work. DO NOT disturb the continuity bond.

On completion of work remove the continuity bond in reverse order:-
First – downstream
Second – upstream

Do not issue electrical certification. All tests carried out only satisfies:-

a) Safe working practice
b) As an aid to combi boiler fault finding procedure

Introduction

Water contained in the central heating system may well be dirty (blackish or brown) which in itself indicates a possible problem with the installation. If this dirty water is splashed onto carpets, walls, furnishings etc. it can stain. This kind of damage is easily avoidable if the necessary precautions have been taken to avoid/catch spillages.

If there is a spillage, deal with it immediately to prevent further damage and possible injury from slipping etc. Wet conditions such as wet hands, feet, floor etc. will reduce the bodies resistance to electric shock *(see Electrical Safety – 4.2)*.

Hot Water

There are two areas of concern with hot water:-

1. Scalding
2. Legionella type infection

Domestic central heating and domestic hot water is about achieving time and temperature comfort levels and the safe storage and delivery of hot water. However, both areas of risk are present and manufacturers do their best to build in controls to minimize the risks. Designers and installers follow regulations which also seek to minimise the risks. Unfortunately, hot water temperatures that do not cause scalding, promote legionella bacteria growth in water systems. Hot water temperatures that kill and prevent legionella bacteria will cause scalding.

Scalding

Do not work with hot water. Allow combi boilers/radiators/hot water pipes etc. to cool before attempting any fault finding and repair. Scalding does happen and yet is wholly avoidable. Temperatures at which scalding can occur are constant, however, the degree of scalding depends on several factors:-

a) The temperature
b) Volume of hot water delivered
c) Contact time

Water need not be boiling to scald. In fact temperatures above 45°C can cause injury quickly. Skin sensitivities vary but as a guide partial thickness burns will occur when hot water is at:-

55°C after approximately 30 seconds
60°C after approximately 5 seconds
70°C at about 1 second

Full thickness burns will occur when hot water is at 70°C after about 10 seconds.

For interest, below is a list of approx. 'safe' water temperatures:-

Bath fill	41–44°C typically 43°C
Shower	40–43°C typically 41°C
Wash hand basin	38–41°C typically 40°C
Bidet	36–38°C typically 37°C

At the kitchen sink hot water needs to be at about 46–48°C to remove grease but there is a risk of scalding.

The dilemma is obvious, hot water over 45°C can scald. Central heating systems use water at much higher temperatures than this. Do not work with installations or appliances that contain hot water. Let it cool.

Thermostatic Mixing Valve (TMV's)

The objective when distributing hot water from a combi boiler is to keep the pipe runs to a minimum (and insulated) and by fitting thermostatic mixing valves close to the draw off points (maximum 2 metres from for point of use) reduce possible scalding.

Thermostatic mixing valve

Thermostatic mixing valves are normally easily fitted into the existing pipe runs under baths, wash hand basins and kitchen sinks. It is fitted in the hot supply with a tee across to the cold supply. The cold then mixes with the hot in the thermostatic mixing valve and exits the draw off point at the required adjusted pre-set temperature. Do not solder pipe work close to the valve body as it contains temperature sensitive components. Also available are thermostatic bath mixer taps. This entails exchanging the old bath/mixer taps for the thermostatic bath mixer. Your wholesaler will be able to advise. Remember to check the tap hole spacing in the bath will take the new mixer. They usually do.

Legionella Type Infection

The pathogen (an organism that causes disease) most commonly associated with the legionella type infection which gives rise to legionnaires disease is a bacterium called Legionella Pneumophila. Surprisingly, this bacterium is a fairly common microbe and is found in small numbers in rivers, lakes and reservoirs. Fortunately, if the bacterium is ingested the prevalence of disease development is extremely low. It is much more dangerous when it is inhaled via a water type spray (aerosol). Once in the lungs it can proliferate leading to a severe pneumonia. Approximately 1 in 8 (12%) of cases be fatal.

Legionella can survive in water temperatures of 20–60°C. Under 20°C or at 60°C and above it dies. Worryingly, the temperature at which it multiplies prolifically is the healthy body temperature of 37°C. It grows fastest if there is a good source of nutriments. It may get its nutriments from biofilm. This is a microbiological layer which develops on the internal surfaces of many water systems and provides a habitat for bacteria.

Decreasing the risk of legionella infection

1. Do not have passive dead legs – lengths of pipe which have no flow through them should be avoided
2. Keep cold water pipe runs to a minimum so the volume of system water is limited
3. Use low corrosion metal pipes e.g. copper is a natural biocide
4. Do not install cold water pipes above hot water pipes in horizontal runs and keep hot and cold water pipes well separated on vertical runs. These pipes should be insulated
5. Central heating water (primary) should NEVER be allowed to mix with the domestic hot water or domestic cold water

Service & Repair

Contents

Introduction

This chapter on service and repair covers the tools and methods that may be used in the service and repair of combi boilers. Conveniently, it also covers such procedures as let-by and tightness testing, purging, standing and working gas pressures along with burner pressure(s). The section on the use of multimeters is unique and this alone will probably be reward enough for many owners of this book. This chapter may also serve as an aide memoir for experienced professionals.

The tools needed for service and repair are arguably the same as for a full central heating installation and possibly the same as a boiler replacement as most installers/engineers need to provide a full service to the customer, although ultimately, it is the professionals choice. This chapter also covers in some detail sections that you may not have a great deal of experience such as flue gas analysers, electric pipe freezers etc. For some this might also include pump tools, weir gauges, mains water pressure gauge and electronic manometers.

If the customer reports low central heating temperatures and all is well with gas pressures and the combi boiler, the fault may well be with say 'radiator balancing' or 'sludged up' radiators. This must also be corrected and the true professional will have the knowledge and the tools to eliminate the cause of the customer complaint *(see 'Central Heating – Fault Finding & Repair' book).*

On the other hand, the professional may decide to concentrate his daily activities on servicing boilers and breakdowns. If this is the case, core drills and the power tools to drive them, ladders, roof ladders, masonry and plastering tools etc. will not form part of the tool kit.

Gratifying and rewarding as it is, it's not everyone who can face a boiler, repair and/or service it, move on to the next boiler and repeat! Most commonly installers and engineers alike, especially one man bands, find variety of work the most pleasurable and the most satisfying. To own the tools and have the knowledge required to confidently repair combi boilers, service them, install them, discuss central heating and hot water requirements with the customer, identify possible problems before work commences and so on is all invaluable to the professional.

Read this chapter and when you decide what your daily work routine is to consist of, buy the tools you require, buy wisely and buy quality. In all likelihood you will need them all with perhaps one exception. Organise a wrist watch with a stop watch facility for a birthday or Christmas present and the 'hand held stop watch' money will buy you a 'pump tool' and some beer!

Flue Gas Analysers

These are rapidly becoming an essential 'piece of kit' for every installer/engineer. Quality analysers have more than just an analysing capacity e.g. differential thermometer *(see Flue Gas Analysers – 5.3)*. At commissioning and in many fault finding situations it is necessary to measure the temperature of domestic hot water at a known flow rate *(see below Weir Gauge)*. The flow and return temperatures must also be measured to enable effective and correct commissioning, system balancing, as a diagnostic strategy etc.

Flue Gas Analyser

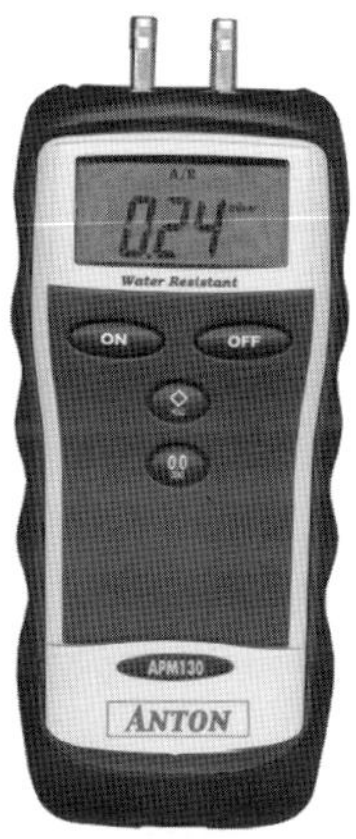

Manometer

Manometers

Water filled or electronic manometers are an essential tool and are used to measure gas pressures (see Manometers – 5.2). Interestingly, a quality flue gas analyser also has the function of single and double pressure measurements. When you own a good flue gas analyser the water filled 'U' gauge will be redundant along with your gas leak detector, carbon monoxide detector and differential thermometer.

Weir Gauges

Also known as a flow meter, flow gauge or flow cup. This is an essential tool used for measuring the 'flow rate' of mains domestic cold water and combi boiler heated domestic hot water. A weir gauge MUST be used when assessing the possibility of installing a combi boiler.

Weir Gauge

The mains cold water draw off point, usually at the kitchen sink, is fully opened with the weir gauge placed in the stream of water. The weir gauge allows water to escape over an invert and is calibrated in litres/minute. A reading is taken and a reading of less than 20 litres/minute giving cause for further investigation *(see Water Flow/Pressure – 2.2)*.

Water pressure and tap designs can give a false impression of flow rates so DO NOT guess. You may only discover the inaccuracy of your guess after the combi boiler has been fitted, commissioned and is not operating correctly! It is possible to use a litre jug and using a stop watch calculate the flow rate; messy, inaccurate and may cause the customer to loose confidence in you. Use a weir gauge for measuring cold water flow rates and for confirming the combi boiler is delivering the litres/minute of hot water the manufactures claim. You will need a weir gauge at site survey, commissioning and for some fault finding situations.

Pressure Gauge

A pressure gauge MUST be used at the initial site survey and will also be used for some fault finding situations. Mains water pressure can be measured at a mains cold water draw off point, usually at the kitchen sink to establish there is at least 1 bar pressure *(see Water Flow/Pressure – 2.2)*.

Many combi boilers will not operate correctly if the mains pressure is extremely low, typically less than 0.5 bar. As mains pressure varies throughout the day, depending on local demand, this measurement must be taken with demand peaks and troughs in mind. It is not possible to guess and remember that flow is not the same as pressure.

When attaching the gauge to the draw off point ALWAYS use the safety chain. Avoid scratching the tap by either passing the chain through a flexible plastic tube or use a cloth between the chain and the tap.

Pressure Gauge

Pump Tool

Pump tool

The circulating pump fitted within most combi boilers usually outlasts the combi boiler unless the central heating system has a built-in fault or develops faults. This can shorten the working life of the pump and eventually it must be replaced *(see Pumps – 6.18)*.

Pump venting using pump tool

Replacing a built-in pump is often a difficult and awkward procedure so on many occasions the body of the pump is left 'in situ' and the head is changed. The bolts/studs used to attach the head to the body are often inaccessible so a 'pump tool' is used. This tool makes a difficult job easy. As the pump MUST be vented at commissioning and when the combi boiler is drained down the pump tool has a vent screw blade in the handle. This blade helps prevent 'burring' of the vent screw slot and gives appropriate leverage for seized vent screws, during the venting procedure. To vent the pump the impellor must be turned a few times and the pump tool has a reversible 4mm blade/hexagonal bar arrangement designed for the purpose. This is an invaluable piece of kit and is a must for all tool boxes. Ask at your local plumbing and heating wholesaler or go on line

Removing pump head using pump tool

Electric Pipe Freezers

Electric pipe freezers are invaluable in many 'wet' plumbing situations. Especially so when working on combi boilers and draining down is necessary. The whole system can be drained down or just the combi boiler drained. Many books/ instructions will simply say 'close isolates on the flow and return pipes and drain down combi boiler…..' This is true for new or relatively new combi boilers and for some quality combi boilers. However, closing the isolating valves fitted to many older combi boilers causes a whole set of problems as they break and must be replaced *(see Pipe Freezers – 5.5)*.

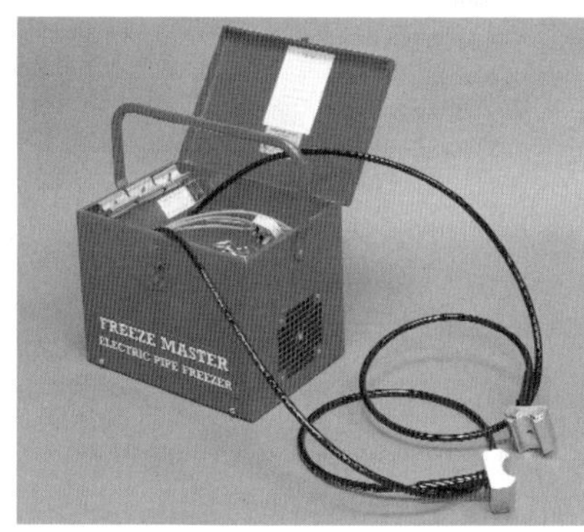

Pipe freezing machine

Power flush machine

Power Flush Machine

It is a requirement that for new central heating installations and for replacement combi boilers, the system is flushed. This will remove installation debris from new installations and sludge/scale from existing system radiators and pipe work. Using a power flush machine meets this requirement and ensures the manufacturers installation instructions and guarantee are not compromised *(see Flushing & Protecting – 2.5)*.

Stop Watch

Hand held or a wrist watch with stop watch facility is essential for timing let-by and tightness tests *(see Let-by Test – 5.6 and Tightness Test – 5.7)*. Also used for measuring active dead leg time lag for water at the required temperature to be delivered at the draw off point *(see Dead Legs – 2.3)*.

Introduction

A manometer (pressure gauge) is also referred to as a water gauge or 'U' gauge. A manometer is a vital piece of equipment and all installers/engineers must be familiar with its correct use. An electronic manometer is also known as a digital gauge. Whatever term is used, water filled or electronic, these instruments measure pressure in millibars i.e. mbar. A millibar is one thousandth of a bar.

Water Pressure Gauge

This instrument is a 'U' shaped tube, originally made of glass but nowadays more commonly plastic, which is approximately half filled with water in both 'U' tube limbs to the zero level on a graduated scale. This movable scale is set behind the 'U' tube limbs and gives limited adjustment to allow for final zeroing. The graduated scale uses millibars as the unit of measurement and for natural gas must have a minimum scale range of 0 – 30 mbars. If used correctly is accurate to within 0.5 mbars.

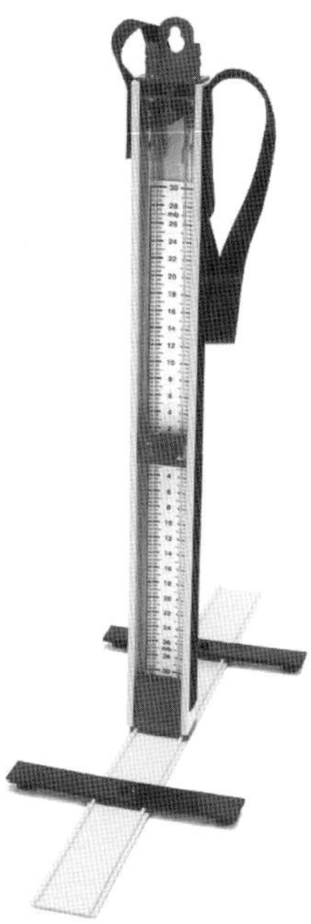

Water filled manometer

The gauge can be connected by a rubber tube to a variety of test points e.g. on the outlet side of the gas meter, at the combi boiler, at the modureg gas valve etc. The gas pressure within the gas pipe work/installation/appliance will cause the water to rise in the 'open to atmosphere' limb of the 'U' gauge.

To ensure the rubber tube end connection to the 'U' gauge is gas tight, occasionally check the connection with leak detection fluid.

When using the 'U' gauge:-

- Check the gauge is in good condition and the glass/plastic is clean
- Using clean water set water levels at zero
- Ensure there are no air bubbles in the water
- Keep the 'U' gauge in vertical position
- Admit gas slowly to prevent a pressure surge blowing the water out the open end
- Read the scale at the base of the meniscus
- Keep eyes in line with water level to avoid parallax i.e. incorrect angle of sight

Note:- If the gauge cannot be zeroed add together both limb values and divide by two. e.g. the gauge reads 17 mbar in one limb and 23 mbar in the other:-

17 + 23 = 40   2 = 20 mbar

Electronic Pressure Gauge

This instrument is easier to use than a 'U' gauge, saves time and is more accurate than a water filled 'U' gauge. Water is not required and zeroing is not necessary other than pressing the zero button before commencing the procedure and checking for zero reading at the end of each test. However, an electronic pressure gauge must be calibrated and serviced annually with a calibration certificate issued which will incur a small fee. Only quality electronic gauges that are intrinsically safe for use in a flammable atmosphere must be used and must be accurate to within 0.3 mbars.

Again, from time to time check the rubber tube connection for gas tightness using leak detection fluid.

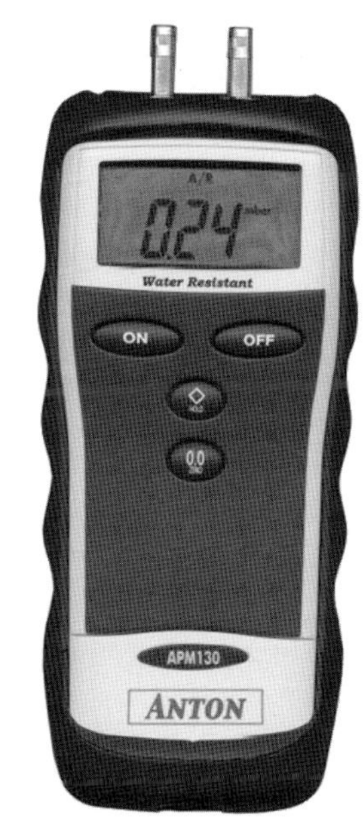

Digital manometer

Manometer Uses

The manometer is an essential 'piece of kit' and will be used for measuring standing pressure, working pressure, burner pressures and for gas tightness tests:-

When a smell of gas is reported
Newly installed meters or pipe work
Before work is carried out on existing pipe work
When work is carried out on a gas fitting/pipe work/combi boiler
On completion of pipe work extension or alteration
Before wrapping, painting, burying or concealing pipe work
During routine testing of existing pipe work
Before restoring the gas supply after work on an installation
Immediately before 'purging'

Digital manometer with hoses

Introduction
Short of completely stripping a combi boiler, a flue gas analyser also known as a combustion meter or portable electronic combustion analyser, provides the most reliable means of checking that everything inside is working correctly. Simple numerical indicators of safe and efficient combustion are displayed and can be read directly from a screen or printed off for future reference, offering reassurance to customers and appliance users as well as the installer. The use of these handheld devices is fast moving from a good idea to a matter of necessity. However, they are not an alternative to commissioning and servicing requirements as indicated in the manufacturer's instructions and always check the flue and flue terminal is satisfactory *(see Flues & Terminals – 3.1)*. The flame picture, for so long considered to be satisfactory if burning with a blue flame (not orange or yellow), is no longer considered 100% reliable. In some instances it is not possible to see the flame.

Some combi boiler manufacturers already insist that flue gas analysers are used in the commissioning and servicing of combi boilers using forced draught premix burners, with failure to comply potentially leading to prosecution for the installer.

Gas Leak Detectors
While not strictly flue gas analysers, simple gas leak detectors used to check the air around an appliance are worth considering here. Easy 'on-off' devices with basic displays indicating gas levels graphically or numerically, they are available in separate versions to detect leaks of flammable gas or carbon monoxide.

To ensure that pockets of gas are not missed when the air is 'swept' with the device, a rising alarm tone indicates an increase in the detected gas concentrations. Instruments like these are typically self-calibrating; turned on in fresh air, they automatically zero themselves within a few seconds.

Flue Gas Analysers

Flue Gas Analyser

Flue gas analysers draw exhaust gases directly from the flue, providing vital information about safety and efficiency. By sampling the gas mixture coming from the appliance and measuring its chemical and physical make-up, multi-function flue gas analysers offer the engineer considerable diagnostic capability.

Ideally, all the carbon in a fuel would be converted by combustion to carbon dioxide (CO_2). In practice, this is impossible: some carbon reacts only partially with the oxygen in the air, leaving a trace of carbon monoxide (CO) in the exhaust gas. The quantity is generally tiny and it should escape harmlessly through the flue. The analyser displays the CO levels in parts per million (PPM). However, a limited or inadequately ventilated air supply will leave too much CO, exposing those nearby to the risk of poisoning. The ratio of carbon monoxide to carbon dioxide ($CO:CO_2$) in the flue is therefore a simple test of safe and efficient combustion, with a low result suggesting a safe and efficient combi boiler.

The following combustion ratio trigger valves are used to assess the combustion performance of a combi boiler:-

Reading greater than 0.004 – Combi boiler requires servicing.

Reading of or less than 0.008 following a service, indicates the combi boiler is operating safely and efficiently even though the reading is above the 0.004 trigger valve.

Reading greater than 0.008 requires additional checks to ensure the combi boiler is assembled, cleaned and operating as per the manufacturers instructions.

5.3 Flue Gas Analysers

When the combi boiler has been serviced if the combustion ratio ($CO:CO_2$) is greater than 0.008 the following MUST be checked:-

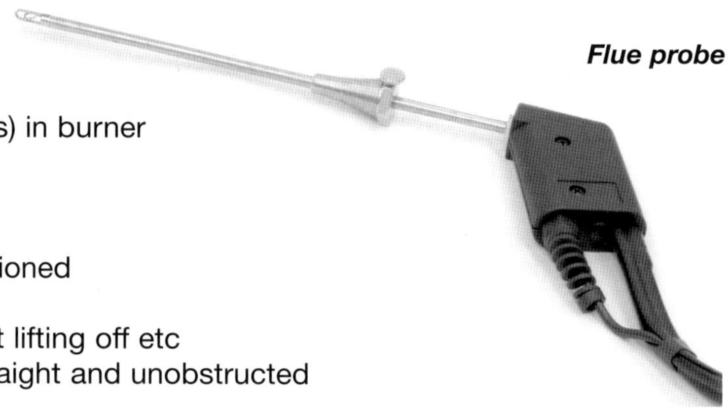

Flue probe

Gas meter operating pressure
Combi boiler working pressure
Combi boiler gas rate at meter dial
Internal gas leak(s) at connections/split(s) in burner
Correct pilot injector size
Correct pilot position
Burner and ports are clean
Burner is correctly assembled and positioned
Burner is undamaged i.e. no dents etc
The flame picture is correct i.e. blue, not lifting off etc
Heat exchanger is clean and fins are straight and unobstructed
Baffles are correctly positioned
Flue and terminal position is correct, joints are sealed and flue gases are not re-circulating causing vitiation i.e. terminal positioned in light wells

Should the reading still be greater than 0.008 when ALL the above have been checked contact the combi boiler manufacturer. DO NOT operate dip switches or turn potentiometer on the printed circuit board without manufacturers instructions.

Handheld flue gas analysers measure the concentrations of both gases and automatically provide the numerical ratio. By indicating the need, if any, for further investigation, this facility can avoid unnecessary and expensive stripping of 'healthy' combi boilers. Good flue gas analysers include a timed test feature to check CO accumulation around the appliance.

Optimising efficiency
It is essential that the correct mixture of air and fuel is available to the burner. As well as a poisoning risk and reduced efficiency, too little air – that is, too little oxygen – creates an unstable flame. If the flame extinguishes it will cause the combi boiler to shut down, possibly to lock out. To avoid these situations, combi boilers are set up to maintain excess air in the flue. However, if the air flows too quickly, significant heat will be lost in the flue, making the boiler inefficient. This is especially true for condensing combi boilers, which can only achieve the high efficiency ratings claimed by the manufacturers if the gas/air mixture is set precisely.

Percentage combustion efficiency is derived mathematically from flue gas temperature and oxygen content. Traditionally calculated with an oxygen monitor, a thermometer and a slide rule-type device for entering variables such as fuel type, efficiency is now worked out instantly by the flue gas analyser. For simplicity, all readings are taken via a single probe. The effect of any burner adjustment is shown in real time, making it easy to fine-tune for optimum efficiency.

Digital manometers
As well as measuring ambient gas levels of both natural gas and carbon monoxide, $CO:CO_2$ ratio and efficiency, flue gas analysers now offer additional features to make the engineer's work easier. Flue gas analysers with built-in digital manometers dispense with the need for a manual pressure gauge, showing negative and positive pressures as well as differential values to indicate the burner pressure drop. Usefully, one analyser allows the pump to continue running while pressure measurements are taken, ensuring a continuous, real-time display of the effect on performance of every adjustment.

Good flue gas analysers offer a unique timed pressure test. Following the Let-by and the Tightness Test, which would normally be carried out with a cumbersome water manometer, it checks the gas-tightness of a system over a two minute period following the installation of a combi boiler, or any work carried out on gas pipe work or gas fittings.

Thermometer

To adjust flow and return temperatures, a differential digital thermometer facility may be included, along with special probes and pipe clamps for various pipe diameters. Not only useful for combi boiler flow and return pipes, the differential thermometer feature is equally applicable to individual radiators, improving their efficiency and balancing the system.

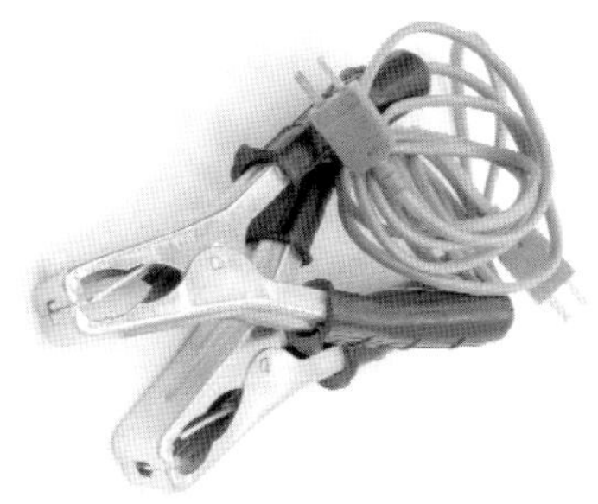

Pipe clamps

Recording data

The diversity of information produced by an analyser needs to be recorded to be useful. A comprehensive internal memory will store all the data within the instrument. Ideally holding a week's records or more, the memory can be downloaded to a PC via a convenient infrared link. Depending on the type of instrument used, an internal or external printer can also be used to create hard copy records. These are especially useful because, when timed, dated, and signed by the recipient, they provide evidence that work was carried out to the required standard. Where offered, a 'live print' function will allow data to be sent directly to the printer without first using space in the instrument's memory.

Above all, it is the 'all-in-one' nature of today's flue gas analysers that makes them a 'must have' tool. Replacing several tools with a single compact instrument has obvious advantages in terms of saving toolbox space. Having only one set of batteries to worry about is also a relief, especially if they are the fast charging nickel metal hydride (NiMH) type, able to take a recharge or trickle charge at any time. Moreover, service costs are dramatically reduced, and if the manufacturer offers a rapid turnaround, annual servicing need take only two or three days.

Using a Flue Gas Analyser – Summary

An instrument capable of the following is invaluable to the installer/engineer and with a protective rubber boot will last many years:-

Performance	CO in ppm, CO_2 in % and CO/CO_2 ratio
Efficiency	O_2 in %, excess air in % and combustion efficiency in %
Carbon Monoxide Leakage	O_2 in % and CO in ppm
Differential Pressure and Tightness Test	High (+ve) and low (-ve) pressure and count up timer with logging of pressure
Gas Leak	Approximate concentrations of flammable gas (levels 0 – 3)
Carbon Monoxide Room Test	Elapsed time in minutes and maximum CO ppm for that minute
Differential Temperature	Flow, return and differential temperatures

Downloadable to a printer and/or computer

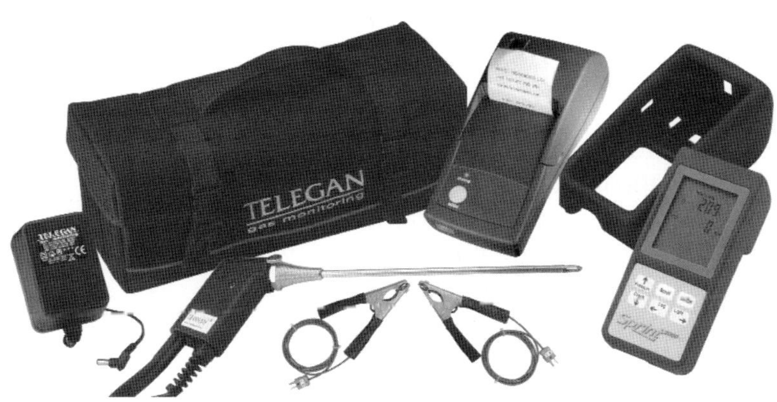

Flue Gas Analyser with protective rubber boot, printer, probe, pipe clamps, charger and robust carry case

5.4 Multimeters

Introduction

Before we look at the correct use of a multimeter read Electrical Safety 4.2 and the following introduction should be especially noted.

Firstly, it is not safe to use a multimeter unless you have an adequate knowledge of electricity and adequate experience of electrical work. You must understand the system to be worked on and have at least some tutored practical experience so you are aware of the hazards which may arise and the precautions you need to undertake during work on that system. If you are unable to recognise that it is safe to continue to work on that system you are at risk of serious injury or even death. Testing mains powered appliances or components is not recommended for beginners. A mistake could be fatal. Consider the following:-

The domestic voltage of 230V is potentially dangerous and when working with more than 25 Volts precautions must be taken as severe burns are likely. Most, if not all, appliances are protected with fuses ranging from 1Amp to 13Amps and at the consumer unit ring mains are often rated at 30Amps. You can feel an electric current at 1mAmp (1/1000th of 1 Amp). At 30mAmp an individual cannot release themselves (due to involuntary muscular contractions) and at 80mAmp (80/1000th of 1 Amp) death is likely. Electricity is dangerous. If you are not competent, do not use a multimeter. Use the services of a central heating engineer for system faults or a qualified electrician for other domestic electrical problems.

Using a Multimeter

Multimeter

The multimeter will carry out several functions measuring, voltage, continuity, resistance and current. Multimeters are available as the traditional analogue type which displays the reading as a needle deflection across a scale and a digital type which displays the reading as numbers. Both types have a top half displaying the reading and a bottom half with the scale selector switch. It is ALWAYS necessary to select the correct scale so that the multimeter is not damaged and the correct reading is obtained.

Do not attempt to use a multimeter unless you are trained and competent to do so. You must not work on gas components/ appliances i.e. the combi boiler unless you are CORGI registered.

Remember – All the following tests are only to allow the engineer to follow safe electrical procedures and for fault finding NOT for certifying the correctness of the electrical installation.

Test 1 – Voltage is either direct current (DC) or alternating current (AC). Direct current is symbolised as 'V...' and usually measured to check/set up components etc. Alternating current is symbolised as 'V~' and used to check mains voltage. Probably the best advice is to avoid tests on the mains supply as far as possible. However, to prove the combi boiler/component has been correctly isolated from the mains supply or has voltage:-

1. Insert the black test lead into the multimeter terminal marked 'Com'
2. Insert the red test lead into the multimeter terminal marked 'VΩ' or 'ACV' or 'V' or '+'. Often there is a lightening symbol with or without an exclamation mark next to the terminals. DO NOT insert either lead into the terminal marked '10A' or '20A' or 'ΩA' or 'mA'
3. Select scale – voltage AC marked as 'V~' or 'ACV'
4. Select range – 500 or 600 or 750. Less than 250 will read off the scale
5. With analogue multimeters zero the indicator
6. Test the multimeter on a known supply. Red to live and black to neutral – it should read approx 230V
7. Conduct AC voltage test on the combi boiler/component. Red to live and black to neutral – it should read zero volts

8. Re-test combi boiler/component
9. Re-test the multimeter on a known supply. Red to live and black to neutral – it should read approx 230V

Remember – Testing live voltage (230V) is potentially hazardous. ALWAYS follow safe working practices.

Test 2 – Continuity testing is to establish whether or not there is a continuous circuit. This reading will prove if a switch or a component with windings is functional. We are only interested in two readings:-

Infinity	-	no continuous circuit
Zero Resistance	-	continuous circuit

Notes:-

a) If the multimeter is set on resistance (Ω) and the probes are touched together there may be a reading due to the resistance of the probes and leads

b) If measuring across the live to neutral parts of a circuit a reading of zero resistance (continuity) would indicate a 'short circuit.' If the combi boiler is turned on or connected to an electrical supply the fuse may blow/trip

c) i) Switches that are closed read zero resistance (continuous circuit)
ii) Switches that are open read infinity (no continuous circuit)
iii) Windings that are functional read a resistance (the value will vary with each type of winding)
iv) Windings that are not functional read either infinity (winding circuit is broken) or zero resistance (the winding has short circuited between live and neutral). DO NOT turn on the electrical supply or the fuse may blow/trip

Continuity checks might be carried out on:-

Components (Windings)	Switches
Solenoid Valve	Fan Thermostat
Fan Unit	Boiler Thermostat
Transformer	Limit Switch

When testing for earth continuity the combi boiler/component MUST be electrically isolated. ALWAYS test for zero voltage reading before carrying out any other procedure.

The multimeter can be set to resistance (Ohm's scale marked Ω) or to continuity marked ·)))

1. Insert the black test lead into the multimeter terminal marked 'COM'
2. Insert the red test lead into the multimeter terminal marked 'VΩ' or 'Ω' or '·)))'. Often there is a lightening symbol with or without an exclamation mark next to the terminals
   DO NOT insert either lead into the terminal marked '10A' or '20A' or 'ΩA' or 'mA'
3. Select scale – Continuity marked ·))) or Ohms marked as Ω
4. Select range – if no dedicated continuity setting choose 20–100Ω
5. With analogue multimeters zero the indicator
6. Test the multimeter on a known resistance i.e. touch probes together. It should read less than 1Ω
7. Conduct continuity test on combi boiler/component. When testing earth continuity on the combi boiler touch the black to the earth connection on the plug or fused connection unit and the red to the metal parts of the combi boiler i.e. casing, pump, fan, collector hood etc. It MUST read less than 1Ω (after subtracting lead resistance)
8. Re-test combi boiler/component
9. Re-test multimeter by touching probes together as in '6'

Test 3 – Resistance of circuits or components can be checked using the resistance scale. When using the resistance scale (measured in ohms Ω) the installation/combi boiler must be electrically isolated otherwise the multimeter fuse will blow and the instrument may be damaged beyond repair. All capacitors in the circuit must be discharged. When working on domestic central heating electrics, testing the resistance is seldom, if ever, carried out. Continuity checks are usually all that is needed.

The short circuit test confirms the integrity of the internal resistance between Live & Neutral of the Combi Boiler. When testing for short circuit the combi boiler MUST be electrically isolated. Remember to first test for voltage zero reading.

1. Insert the black test lead into the multimeter terminal marked 'COM'
2. Insert the red test lead into the multimeter terminal marked 'VΩ' or 'Ω' or '·)))'. Often there is a lightening symbol with or without an exclamation mark next to the terminals.
   DO NOT insert either lead into the terminal marked '10A' or '20A' or 'ΩA' or 'mA'
3. Select scale – Continuity marked ·))) or Ohms marked as Ω
4. Select range – if no dedicated continuity setting choose 20–100Ω
5. With analogue multimeters zero the indicator
6. Test the multimeter on a known resistance i.e. touch probes together. It should read less than 1Ω
7. Conduct short circuit test on combi boiler. Touch the black to neutral and the red to live. A fuse of 3 Amp must give a reading of at least 100Ω. Less than 100Ω indicates a short circuit condition. A 5 Amp fuse must read at least 60Ω. DO NOT switch on the combi boiler. A reading of Mega Ohms (MΩ), overload (OL) or infinity (∞) confirms the integrity of the internal resistance i.e. there is no short circuit
8. Re-test the combi boiler
9. Re-test the multimeter by touching the probes together as in 6

Test 4 – Resistance to Earth test confirms the integrity (resistance) of the insulation between Live & Earth of the combi boiler. When testing for the resistance to earth the combi boiler MUST be electrically isolated. Remember to first test for voltage zero reading.

1. Insert the black test lead into the multimeter terminal marked 'COM'
2. Insert the red test lead into the multimeter terminal marked 'VΩ' or 'ACV' or 'V~' or '+'. Often there is a lightening symbol with or without an exclamation mark next to the terminals.
   DO NOT insert either lead into the terminal marked '10A' or '20A' or 'ΩA' or 'mA'
3. Select scale – Continuity marked ·))) or Ohms marked as Ω
4. Select range – if no dedicated continuity setting choose 20–100Ω
5. With analogue multimeters zero the indicator
6. Test the multimeter on a known resistance i.e. touch probes together. It should read less than 1Ω
7. Conduct the resistance to earth test on combi boiler. Touch the black to the main earth terminal and the red to all the live connections. A reading at lease than 1 Mega Ohm (MΩ), overload (OL) or Infinity (∞) confirms the integrity of the insulation between Live & Earth

Test 5 – Polarity testing will confirm (or not!) that the correct polarity of the mains is applied to the combi boiler.
Caution:- This is a live test. It will prove the live, neutral and earth are connected correctly.

Caution:- With the mains connected, readings are taken across the live, neutral and earth terminals at the combi boiler mains supply wiring harness i.e. where the connecting cable from the 3 pin socket or fused connection unit connects to the combi boiler test lead.

1. Insert the black test lead into the multimeter terminal marked 'Com'
2. Insert the red test lead into the multimeter terminal marked 'VΩ' or 'ACV' or 'V~' or '+'. Often there is a lightening symbol with or without an exclamation mark next to the terminals.
   DO NOT insert either probe into the terminal marked '10A' or '20A' or 'ΩA' or 'mA'
3. Select scale – voltage AC marked as 'V~' or 'ACV'
4. Select range – 500 or 600 or 750. Less than 250 will read off the scale

5. With analogue multimeters zero the indicator
6. Test the multimeter on a known supply. Red to live and black to neutral. It should read approximately 230V
7. Conduct AC voltage test on the boiler mains supply wiring centre:-
   Live – Neutral should read approximately – 230V
   Live – Earth should read approximately – 230V
   Neutral – Earth should read approximately – zero to 15V
8. Re-test combi boiler as in 7
9. Re-test the multi meter on a known supply. Red to live and black to neutral. It should read approximately 230V

Any other reading would indicate a polarity fault and the boiler should be electrically isolated and the live and neutral wires exchanged at the terminals. A reading of 115V across neutral and earth and 115V across live to earth would indicate any open circuit. Ensure the domestic circuit is not totally reversed! If it is, use the services of a NICEIC Registered or competent electrician.

Current is measured in Amperes or Amps symbolised as ‘A‘. It is used mainly to check very small currents in the range of 1–10 micro Amps (µA). Many basic multimeters are unable to read such small currents. Professional gas and heating engineers may have very sensitive multimeters capable of reading those small currents (especially to check flame rectification circuits) and will always place the meter in series with the current path.

In summary:-

Appliance electrically isolated

Test 1 – Voltage check	Multimeter on V ac-Test across L & N – zero reading – O.K
Test 2 – Mains earth continuity	Multimeter on Ω touch earth and metal parts – less than 1Ω - O.K
Test 3 – Short circuit	Multimeter on Ω - Test across L & N more than 100Ω - O.K
Test 4 – Resistance to earth	Multimeter on Ω - Test across L & E reading greater than 1 MΩ - O.K
Test 5 – Polarity	Multimeter on V ac – Test across L & N – 230V O.K., L & E-230V O.K., N & E 15V – O.K

Do not issue the householder with certification of Electrical Wiring

Socket Tester

Socket testers will check for wiring errors including reverse polarity and check the presence of an earth. Others test the earth loop impedance checking within selected bands without tripping the R.C.D. Some also test voltage is low/high (195V/270V) or below 160V.

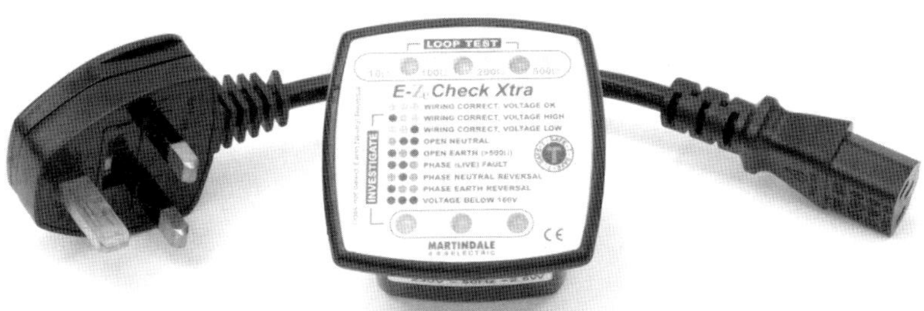

Socket tester with adapter

Introduction

There are many occasions when it will be necessary to fit or repair a wet plumbing component in a combi boiler or central heating installation. Unfortunately, this will often mean having to drain down the combi boiler and/or central heating system with the resultant loss of the central heating system chemicals (inhibiters, anti freeze etc) which have to be replaced when refilling *(see Flushing and Protecting – 2.5)*. Added to the cost of the new or replacement components for the system, you are now faced with the cost of the chemicals. Further, the new fresh aerated water going into the system now releases oxygen, adding to corrosion problems and the system requires venting. In hard water areas more scale will be laid down reducing the efficient working life of the heat exchanger in the boiler and most other wet components within the installation.

Isolation valves to circulating pumps or the combi boiler may not completely stop water flow creating major problems for pump or combi boiler repair. Fitting thermostatic radiator valves to radiators or a weeping radiator valve pipe union will inevitably require the system to be drained. Adding an extra radiator in an extension/loft conversion or changing a panel radiator in the bathroom for a more pleasing towel radiator will also require draining the system. There are many other circumstances that would require a system to be drained. Each requiring several hours to drain, carry out the work and refill unless another method is used. One way is to isolate the working area by freezing the pipes.

The tedious, time consuming and costly drain and refill procedure can be avoided by freezing the pipes to isolate a faulty fitting, allowing removal, servicing or replacement of the fitting in a fraction of the time. The fitting may be a system or combi boiler component.

Each combi boiler will have isolating valves installed at the inlet or outlet for gas, mains cold water, central heating flow and central heating return. There must not be isolating valves installed on the domestic hot water or the pressure relief valve outlet.

The gas and mains cold water inlet isolating valves are usually trouble free and do not let-by. However, on many combi boilers the isolating valves on the central heating flow and return can be troublesome. When draining down a combi boiler it makes sense to close the flow and return isolators. BEWARE, the heated system water can cause these valves to fail. If, when turning them off they are stiff or there is resistance – DO NOT FORCE. They will break and wholesalers may not be able to supply replacements for days or even weeks. In situations like this or on older combi boilers where you suspect the valves, use a pipe freezer.

There are several pipe freezing systems available. One process involves wrapping the pipe(s) with a jacket either side of the pipe area to be worked on and injecting a can of freezing spray via a hose to each jacket. There are jackets for various pipe sizes i.e. 8–15mm, 15–22mm, 28–35mm etc. It is relatively easy to use and when the can is empty a replacement can is purchased. For several freezing jobs, large pipes or for a prolonged job it can start to get expensive and if the spray can runs out before the job is completed there are obvious implications!

Pipe Freezing Machines

Pipe freezing machine

At a very reasonable cost it is possible to hire (from most good hire centres) or purchase a freezing machine. These robust, portable machines will quickly freeze copper, iron, lead or plastic pipes usually within 5–15 minutes depending on the size of pipe. They are equipped with two freeze heads which contain a freeze profile to which sized pipe fittings (up to 88mm!) are attached for a positive flush fit. They come complete with operating instructions, protective gloves, clamps and a mains cable with plug attached.

When it is necessary to drain down the combi boiler use the pipe freezer on the central heating flow and return pipes which in most cases will be either 22mm or 28mm.

With the combi boiler turned off and the system water cold, position the freeze heads at least 200mm away from the isolating valves. If you do this first by the time the electrical safety checks have been carried out, the casing been removed and the coffee is ready the ice plugs will have formed and the combi boiler can be drained down leaving the system water undisturbed.

Freezer head

When a standard freeze head is attached to a warm pipe and the other head to a cooler pipe, pressure within the freeze head is greater at the warm pipe. The refrigerant will tend to precipitate in the head attached to the coolest pipe which causes a freeze delay of the warm pipe. High quality pipe freezing machines with an 'Active Capillary' overcome this problem by delivering pre-programmed quantities of refrigerant to the freeze heads. This is perfect for the flow and return pipes of a central heating system that is cooling down and leaves the system water undisturbed.

Extremely easy to use they are fast, clean and efficient. Once plugged into the mains the freeze will continue indefinitely so the machine can be left unattended allowing other work to be done. The ingenious design of the miniaturised freeze heads and clamps means that pipes in the most awkward recesses (or even fitted flush to a wall) can be frozen quickly and easily. The refrigerants are ozone friendly and its operating parameters are calibrated to allow colder mean temperatures with greater sensitivity to unwanted heat pick up. This micro processor control allows for warmer ambient temperature or even heat from a LPG blow torch.

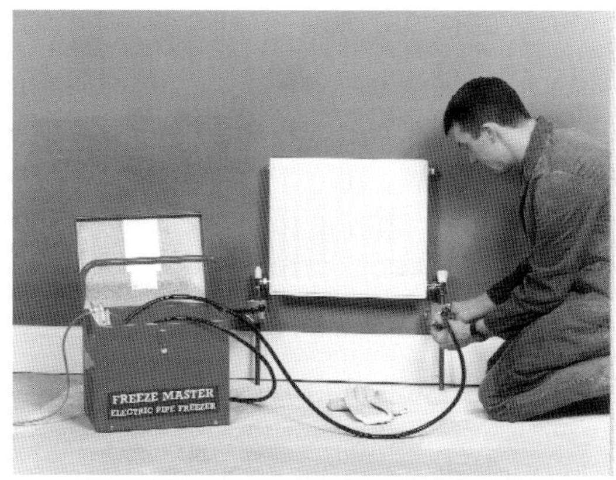

Easy to use pipe freezing machine

Note – Freezing temperatures of -35°C are easily obtained with temperatures as low as -53°C having been recorded.

DO NOT allow bare flesh to contact the freeze area. ALWAYS use the gloves provided.

5.6 Let-by Test

Introduction

Before carrying out a tightness test *(see Tightness Test – 5.7)*, a test must be carried out to confirm the meter control valve is not letting gas pass into the installation pipe work. This test is known as a 'Let-by' test.

The test MUST be carried out at 50% of the operating pressure of the installation. As virtually all domestic gas installations operate at 20 mbar gas pressure the Let-by test pressure is 10 mbar. Whilst this 10 mbar pressure is the stipulated test pressure it is often difficult to adjust the meter control valve to achieve this exact pressure. Therefore, as the purpose of the test is to confirm the valve is not letting by, a test pressure of between 9–13 mbar is often used but remember the target test pressure is 10 mbar. At times of high national gas use the mains pressure may drop significantly so DO NOT conduct let-by procedure with a test pressure greater than 13 mbar.

Let-by Test Procedure

Before conducting a let-by test check the installation pipe work for the following:-

Correct support clipping
Correct sleeving through walls
Open ends
Joint fittings for apparent soldering integrity
Apparent correct pipe sizing
The meter is fitted correctly, is stable and secure

For new installation pipe work disconnect gas appliances. However, for existing installations leave appliances connected and ensure safety shut off valves are in the open position e.g. cooker lids are in the upright position.

Note:- Before extending an existing installation ALWAYS carry out a let-by and tightness test on the installation. When all extension works have been completed the complete installation MUST be re-tested.

1. Check all appliances are turned off including pilot lights
2. With the meter control valve turned off attach a manometer to the meter test point
3. Slowly open the meter control valve to raise the pressure to 10 mbar
4. Turn off the meter control valve
5. After 1 minute observe the manometer reading
6. (a) If there is no movement of the manometer reading or any increase is less than 0.5 mbar the meter control valve is deemed to be not letting by
6. (b) If the manometer reading rises more than 0.5 mbar then let-by is suspected. Let-by can be confirmed by attaching a temporary continuity bond *(see Electrical Safety – 4.2)*, disconnecting the meter control valve outlet union and testing the valve with leak detection fluid. If let-by is confirmed the gas supplier must be informed Tel: 0800 111 999 (N. Ireland 0800 002 001)

The let-by test is a preliminary procedure to the tightness test so…*see Tightness Test – 5.7.*

Introduction
The testing of domestic gas installations and following the correct testing procedure is vital to ensure gas safety. The procedures are appropriate where the following are true:-

A gas meter with a volume not exceeding 6m³/hr
An operating pressure not exceeding 21 mbar at the outlet of the meter regulator
Gas pipe work diameter not exceeding 28mm copper DN25 steel and 1 inch diameter lead

A tightness test, pre-checked by a let-by test, would normally be carried out when:-

Installing/replacing/altering or extending gas pipe work
OR
When there has been a complete loss of installation pressure
OR
Immediately prior to purging
OR
As a routine test
OR
There is a suspected or reported gas escape

Tightness Test Procedure
By using a manometer to measure pressure, a tightness test establishes the degree of pressure loss, if any, indicating leakage from the installation. The procedure used depends on the installation being:-

New or existing
Appliances connected or not
Low or medium pressure fed

Note:- The procedure for medium pressure fed installations with or without a test valve is not covered in this section. Please refer to the gas industry competent person/guidance/instructions.

Following a satisfactory Let-by Test:-

1. Slowly open the meter control valve to raise the pressure to 20 mbar
2. Turn off the meter control valve
3. Allow 1 minute to pass to allow for temperature stabilisation
4. Observe the manometer reading for a further 2 minutes

New Installations
With new installations conduct a let-by and tightness test with appliances disconnected. The pressure drop must be no greater than 0.5 mbar for a water 'U' gauge and no greater than 0.3 mbar for a digital manometer. With no perceptible drop in pressure and no smell of gas the installation is gas tight.

Reconnect all appliances with any safety shut off valves in the open position e.g. cooker lids in upright position. Conduct a tightness test as before for the whole installation at 20 mbar for one minute and a two minute test period. Again, no perceptible pressure drop is allowed with no smell of gas. If there is a drop in pressure then the gas leak must be at the appliance(s) connector or the appliance itself. Disconnect each in turn and conduct a tightness test to establish the position of the leak or use a gas leak detector. Remember to purge the installation pipe work and all appliances – *see below*.

Existing Installations
The tightness test procedure for existing installations is the same as above, however, there is a permitted leakage rate as shown:-

Meter E6	– 8 mbar over 2 minutes
Meter U6/G4	– 4 mbar over 2 minutes

5.7 Tightness Test

Provided :- **There is no smell of gas.**
Note:- If no appliances are connected or there is a smell of gas than a gas leakage rate is not permitted.

A gas leakage rate outside of the above limits OR when there is a smell of gas the installation must be made safe or the gas leak traced and repaired.

Purging
When work is carried out in relation to a gas meter or gas pipe work it is vital to ensure the meter, pipe work and fittings are purged of air.

It is generally accepted that in a domestic property with U6, G4 or E6 gas meter i.e. a meter capable of passing 6m³/hr of gas and with installation pipe work no greater than 28mm the meter and pipe work can be purged into the building provided the following safety precautions are observed. During the purging operation ensure:-

The customer is made aware of the procedure and there may be a smell of gas
Gas does not accumulate in a confined space
No electrical switch or appliance is to be operated
No smoking
No naked lights
Adequate ventilation is provided by opening doors and windows

To enable the correct purge volume to be calculated the capacity of the meter must be known. This is usually found on the test and registration dial panel but may be found stamped on the front of the meter. The purge volumes are:-

Imperial Meter U6	5 x badge capacity, normally 0.071 ft^3/hr $\therefore$ 5 x 0.0.71 = 0.355ft^3
Metric Meter E6 or G4	0.01M^3

Purge Procedure

1. Inform customer that purging of the gas supply is to commence and no light switches or appliances are to operated, no smoking and no naked flames
2. Establish purge volume i.e. 0.35ft^3 or 0.01M^3
3. Open doors/windows to purge ventilation area. Try to create through draught
4. Turn off all appliances, including pilot lights, and ensure meter control valve is turned off
5. With the meter control valve turned off note position of test dial
6. Select a gas point furthest from the gas meter and open this gas point
7. After a short time turn off the purge gas point
8. Check the gas meter dial and note volume of gas that has passed
9. Repeat opening, closing and checking of dial procedure until the correct volume has passed
10. When the purge is completed close and if necessary test purge point for gas tightness
11. Light ALL appliances in turn, ensuring correct operation and flame picture

Note:- When an installation has been purged, the operative will have ‘worked’ at each appliance. Therefore a ‘duty of care’ exists with the operator to ensure that each appliance is working satisfactorily. A visual check MUST be carried out at each appliance and must include at least:-

Signs of spillage
Ventilation where necessary
Flue
Flame picture
General condition of appliance

Introduction

Gas must be supplied to a combi boiler at the pressure specified by the manufacturer so that the desired heat output and efficiency of the boiler is achieved. It is therefore necessary to have a method of measuring gas rates and a procedure for adjusting burner pressures correctly. When commissioning and fault finding a combi boiler, measuring gas rates and checking/setting burner pressures is essential. Further, when a gas installation is subjected to maximum load, a maximum pressure drop of 1 mbar is permitted *(see Gas Pipe Sizing – 2.1)*. The 20 mbar supplied to each appliance may be reduced within the appliance via pressure regulators, to allow a range of pressures at the burner and therefore flexible heat outputs.

Gas Rates

The gas rate depends on the correct pressure AND on the correct 'set up' of the combi boiler. An over/under sized or partially blocked/worn injector would not change the burner pressure but it could affect the heat output of the combi boiler.

The gas rate test measures the volume of gas being burnt by the combi boiler in M^3/hr. The simple calculation converts the time it takes to pass a known volume of gas through a gas meter to the combi boiler into the heat input in kW. This figure is then compared to the manufacturers value in kW found in the manufacturers' instructions and on the data plate.

On older combi boilers the heat input is given as a gross figure in kW which includes the latent heat lost in the flue gases. On newer combi boilers this latent heat loss is excluded and is given as a net figure in kW. To convert gross to net divide the gross heat input of the combi boiler by 1.11 e.g. a combi boiler with gross heat input rating of 28 kW has a net heat input of 25.23 kW.

$$28 \div 1.11 = 25.23 \text{ kW}$$

Note:- The calorific value of natural gas is 1040btu/ft^3 gross and is the quantity of heat produced when a cubic foot of natural gas is completely burnt in excess air.

To convert btu's to kilowatts divide by 3412
In 1m^3 of natural gas there is 10.76 kilowatts

Rating Procedure With U6 Meter (ft^3)

1. Turn on the combi boiler and turn off all other gas appliances
2. Turn on a hot water draw off point to maximum flow, at the bath is best. This maximum flow will ensure the combi boiler operates at maximum output
3. Using a stop watch, record the time it takes for the meter dial to pass 1 cubic foot of gas – normally one complete revolution
4. Turn off the hot water draw off point

Calculation – use the formula:-

$$\frac{\text{Seconds in one hour x Calorific value of natural gas}}{\text{Time for one revolution of meter dial in seconds}}$$

Example:- Combi boiler data plate has maximum heat input 27.5 kW gross
Time for one revolution of meter dial say 40 seconds

$$\frac{3600 \times 1040}{40} = \frac{3744000}{40} = 93{,}600\text{btu/hr}$$

To convert 93,600btu/hr to kW divide by 3412

$$\frac{93{,}600}{3412} = 27.4 \text{ kW gross}$$

27.4 kW is within 1% of 27.5 kW and would be acceptable.

5.8 Gas Pressures

Gas Rating Procedure With E6 Meter (M^3)

1. Turn on the combi boiler and turn off all other gas appliances
2. Turn on a hot water draw off point to maximum flow, at the bath is best. This maximum flow will ensure the combi boiler operates at maximum output
3. Using a stop watch, record the meter reading and record the meter reading after 2 minutes plus the number of seconds until the next digit
4. Turn off the hot water draw off point

Calculation – use the formula:-

$$kW = \frac{\text{seconds in one hour x cubic metres } (M^3) \text{ x number of kW}/1M^3}{\text{Seconds in 2 minutes plus seconds to next digit}}$$

Therefore

$$\frac{3600 \times M^3 \times 10.76}{\text{Seconds in 2 minutes plus seconds to next digit}}$$

Example:-

First reading 00143.032
Second reading 00143.119
Gas used in timed period is $0.087M^3$

$$\frac{3600 \times 0.087 \times 10.76}{122} = 27.6 \text{ kW gross}$$

Burner Pressures

Various gas pressures are referred to and often differing expressions for the same pressures are used so:-

Standing Pressure – Occasionally referred to as Static Pressure, this pressure test ONLY confirms the presence of gas within an installation. With the meter control valve turned on and all the appliances turned off a manometer attached to the test point on the outlet side of the meter would ONLY measure the standing pressure, attaching the manometer to the combi boiler inlet test point would again ONLY measure the standing pressure.

Working Pressure – Occasionally referred to as Dynamic Pressure, this pressure test measures the gas pressure when the appliances are working. With the meter control valve turned on and either:-

The only appliance installed turned fully on
Or
At least three cooker hotplate burners turned fully on

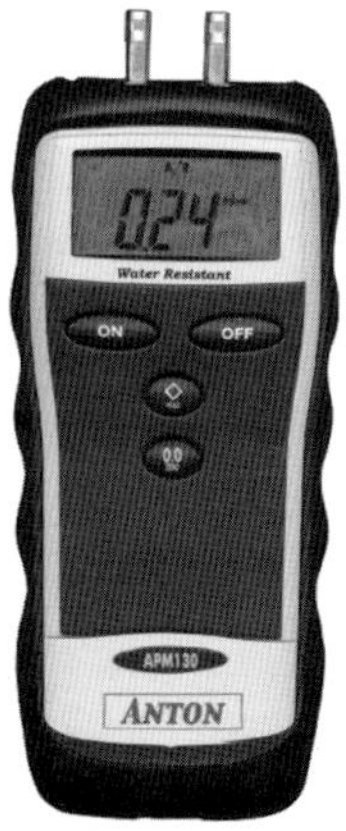

Manometer

A manometer attached to the test point on the outlet side of the meter would measure the working pressure. The working pressure MUST be 21 mbar ± 1 i.e. a maximum of 22 mbar and a minimum of 20 mbar. If the working pressure at the meter is correct with 1 appliance fully on but drops when all the appliances are on then the meter regulator is faulty or the gas service pipe is partially blocked or the meter is undersized for the appliances total gas rate. DO NOT attempt to adjust the meter regulator but report the fault to the gas supplier.

Burner Pressure – Occasionally referred to as Operating Pressure the burner pressure is the pressure of gas immediately behind the injector(s). Therefore, this pressure determines the amount of gas entering the burner mixing tube and the entrainment of air at the burner primary air port. On most combi boilers the burner pressure test point is located on the outlet side of the gas control system, usually on the multi functioning gas valve. The manufacturers instructions will always give the precise location of the burner pressure test point and also the pressure value.

Note:- DO NOT use the burner pressure to assess the rated heat input of a combi boiler. ALWAYS calculate the rated heat input by using the inlet and measuring the gas rate at the meter.

Combination boilers have modulating burners that allow full gas rate for hot water and modulating for central heating, although quality combi boilers also have modulating gas rates for hot water. The burner pressure is controlled electronically by the flow and temperate sensors via a printed circuit board. However, it may be necessary to adjust the maximum and minimum settings by adjusting the brass and plastic nuts on the modureg.

Setting Maximum & Minimum Burner Pressures

1. Electrically isolate the combi boiler
2. Connect the manometer to the burner test point
3. Remove leads from the modureg
4. Restore electrical supply
5. Create a hot water demand by fully opening a hot water draw off point, at the bath is best
6. Check the minimum burner pressure, holding the plastic nut steady adjust the brass nut if necessary, clockwise to increase and anti clockwise to decrease pressure
7. Turn off the hot water draw off point
8. Turn off and electrically isolate the combi boiler
9. Replace leads to the modureg
10. Restore electrical supply
11. Create a hot water demand by fully opening a hot water draw off point, at the bath is best
12. Check the maximum burner pressure, adjust plastic nut if necessary whilst holding the brass nut steady. Always follow the manufacturers' instructions
13. Remove the manometer and replace the test point screw
14. Check the test point for gas leak using LDF

Faults

Excessive Burner Pressure

Reason	Repair
Incorrect setting of modureg	Adjust modureg
Faulty meter regulator	Check working pressure at the meter is correct

Note:- Excessive burner pressure can cause incomplete combustion, appliance overheating and excessive flue gas which the flue system may not be able to clear. If after checking the working pressure at the meter it is due to a faulty regulator the gas supplier must immediately be informed. Treat the appliance(s) as 'immediately dangerous.'

Inadequate Burner Pressure

Reason	Repair
Inadequate pipe sizing	Check pipe sizing and correct if necessary – *2.1*
Damaged pipe work	Check pipe and treat as undersize pipe work – *2.1*
Partial blockage in pipe work	Check each pipe work section in turn and replace pipe work – *2.1*
Faulty meter regulator	Check working pressure at the meter is correct – *see above*

5.9 Servicing

Introduction
It is a regulation that immediately after working on a gas appliance the following will be examined:-

The effectiveness of any flue
The supply of combustion air
The operating pressure or heat input or where necessary both
The safe functioning of the appliance

Further:-

A tightness test MUST be carried out if a smell of gas has been reported
The supply of ventilation air for ALL appliances is adequate
There are no signs of incomplete combustion or signs of spillage from ANY appliance
No appliance appears to be in a dangerous condition

Servicing Procedure - Room Sealed Combi Boilers
The procedure for servicing a combi boiler will vary but a general approach following safe working practices will suffice in the absence of manufacturers instructions. A typical sequence is listed below. Remember,

Always follow the Manufacturers Instructions

Ask the householder if they are experiencing any problems with the combi boiler, domestic hot water or the central heating system

Check for any damage to the combi boiler and inform householder before commencing service

Check the installation of the combi boiler conforms to current gas, water and electrical regulations and best practice standards

Check the combi boiler is level and stable

Check the casing clearances from combustible materials meet the manufacturers instructions

Check the flue condition throughout its length and check flue joints and seals for visual Integrity – *3.1*

Check the flue terminal is correct and a guard is fitted where appropriate and is unobstructed – *3.1*

Check for main equipotential bonding at or within 600mm of gas meter. If the bonding is not present inform the householder to use a competent electrical engineer AND put this advice in writing

Check the gas pipe work for corrosion/sleeving and relative position of electric cables and sockets – *2.1*

Check the system pressure at the pressure gauge – *6.17*

Isolate the combi boiler from the electrical supply and carry out electrical safety tests – *3.4, 4.2, 5.4*

Re-connect and turn on the combi boiler

Check the ignition system operates smoothly and correctly – *6.14*

Check the controls are working correctly

Check the burner flame picture – *6.13*

Check the working pressure – *5.8*

Check the combustion performance using a flue gas analyser – *5.3*

Check the operation of flame supervision devices, maximum 60 seconds – *6.9*

Check external controls are working correctly

Check the rise in system pressure at the pressure gauge – *6.17*

Check the domestic hot water flow rate using weir gauge – *2.2, 5.1*

Isolate the combi boiler from the gas and electricity supplies – *3.3, 4.2, 5.4*

Remove outer casing and clean the main heat exchanger, using appropriate brushes or Tool, taking care not to distort the fins/blades – *6.13*

Dismantle and clean the burner, including any integral filter or lint arrestor – *6.13*

Check the burner for cracks and remove dust and lint from primary air ports and venturi with a soft brush. Remove any blockage with a non-metallic brush – *6.13*

Clean the main burner injector(s) with a soft brush and replace if it appears damaged – *6.13*

Remove the pilot assembly and clean the pilot burner and injector – *6.14*

Check and clean the pilot tube checking for obstructions – *6.14*

Check condition of ignition leads and electrodes and check the electrodes are correctly positioned and connections are secure – *6.14*

Check and clean the flue fan remove dust and fluff with a soft brush. Clean and remove dust and deposits from within the combustion chamber using a soft brush – *6.11*

Replace chamber insulation pads if there is any sign of deterioration

Check fuse rating and visually inspect wiring and electrical connections

Re-establish gas and electrical supply

Check all joints for gas tightness using LDF

Carry out gas purge procedure as appropriate – *5.7*

Check for water leaks

Check system pressure is correct and disconnect filling loop – *6.17, 6.8*

Turn on the combi boiler and check for smooth and correct ignition – *6.14*

Check the burner flame picture – *6.13*

Check the burner pressures are correct and check the gas rate – see manufacturers instructions or the data badge – *5.8*

Check combi boiler controls and external controls function correctly

5.9 Servicing

Check the FSD operates within 60 seconds – *6.9*

Check overheat stat – *6.21*

Check boiler stat – *6.21*

Check flow and return temperatures, 82°C flow and 72°C return, condensing combi boiler, 75° flow and 55° return *(see 'Cental Heating – Fault Finding & Repair' book)*

Check the combustion performance using a flue gas analyser – *5.3*

Return the manufacturers instructions to the customer and leave a copy if you have a spare

Check that the customer understands how to use the combi boiler controls

Remind the customer the combi boiler should be serviced at 12 monthly intervals

Leave the work area clean and tidy

Components

Contents

Introduction

This chapter deals with the major components contained within combi boilers. With the huge number of combi boilers installed (about 10 million in the U.K.) and the variations that can occur, major or minor, it is impossible to specifically describe each and every component. Fortunately, ALL components carry out definable actions and in most instances will share basic functions with its counterpart in other combi boilers. It is important to have enough knowledge to enable an understanding of each component, its action and how it relates to other components within a combi boiler.

Each section of this chapter reviews each major component with an:-

Introduction – Giving an overview of that component, various names it may be know by, variations that may/do occur, its function within the system and relationship to other components.

Design and Construction – This looks at various physical properties of the component such as manufactured in metal/plastic, operating temperature/speeds/pressure, testing and quality assurance standards, position, connection/mounting methods, safety and regulation issues.

Operation – This takes into account the 'variations on a theme' for that component and describes how the component is activated/deactivated. This also considers the input, processing and output of signals/activating mechanisms, mode of operation and its relationship to other components.

Fault Finding – This describes what can fail with the component, what can cause the failure, what faults this component may develop, its impact on the operation sequence of the combi boiler, signs of impending failure, why this may happen and the fault logic that can be used to detect the faulty component. The fault logic charts give a clear and concise 'Fault – Reason – Repair' summary enabling a full and complete picture to be built as to why the combi boiler performance is impaired.

Service and Repair – Many components are 'fit and forget' and require no servicing although it is always wise to check for impending component failure whenever possible. Manufacturers instructions give detailed service procedures but most servicing can be carried out with a good comprehensive knowledge of that 'type' of combi boiler. Having established where the fault is and which component(s) is causing the interruption of boiler sequence.

The repair of that component(s) often, but not always, involves replacement on a like for like basis and is straightforward in most cases. This repair/replacement is described and also any further action that should be taken to complete the job.

Introduction

With many modern combi boilers using a fan to assist in the removal of the products of combustion (flue gases), it is imperative that the combi boiler does not operate if the fan is not running or is ineffective. Therefore a gas control and safety device must be used to assess that there is an adequate flow of air/flue gases. This 'flow' switch (there are 'flow' switches on the primary and secondary water) is an air flow proving device and is commonly called an 'air pressure switch.' It is also known as a 'fan pressure switch,' fan proving switch' or a 'differential pressure switch.' Whatever term is used it is the safety device which proves the effective operation of the fan prior to the ignition process. Further, if the fan fails at any time (including slowing down) this is detected by the air pressure switch and a sequence of events will follow which ends with the gas valve closing onto its seating and the combi boiler shutting down.

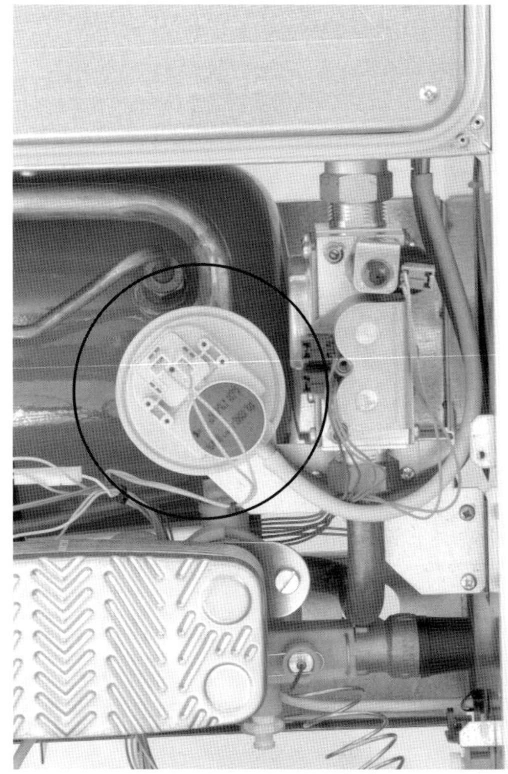

Air pressure switch

Many modern forced flue combi boilers use thermistors *(see Thermistors – 6.20)* instead of an air pressure switch and air pressure hoses. There are usually 2 and they measure the temperature of the air at two different locations within the air and flue ways. The differential temperature sensing is transmitted to the printed circuit board and will prove:-

(a) the fan is running to speed
(b) the flue ways are unobstructed

Design and Construction

Polycarbonate and metal air pressure switch designs

Situated outside and either above or below the combustion chamber the electro-mechanical air pressure switch consists of a flat, air tight capsule made of polycarbonate or metal, internally divided into 2 halves by a thin and delicate NBR-based rubber or silicon diaphragm. (Blowing into an air pressure switch can damage the diaphragm!) Each side of the diaphragm there is a pressure connection, one for positive fan pressure and the other for negative.

There is a micro switch mounted on the negative side. The air pressure changes created by the fan are transmitted to the air pressure switch by flexible tubes often referred to as air pressure hoses or impulse tubes. They are attached to the pressure connection ports. Often these connectors are marked ±, +ve/-ve or 1 and 2 where 1 is positive and 2 is negative.

Some air pressure switches are one port type but most have two.

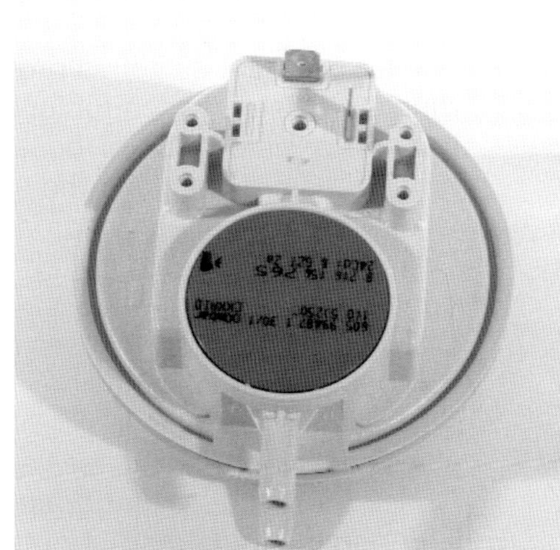

Alternative air pressure switch design

6.1 Air Pressure Switches

Another popular air pressure switch design

Operation
When the fan is running a pressure differential is created and transmitted via the air pressure hoses either side of the diaphragm which causes the diaphragm to deflect to the negative side. The diaphragm moves a spindle which operates a micro-switch sending a signal to the printed circuit board to commence the ignition sequence and the gas valve opens.

On older fan assisted combi boilers with a permanent pilot flame the air pressure switch has two stages. The slower fan speed is to maintain the pilot flame and high speed for the main burner. With the pilot flame alight and the fan running at slow speed the switch is in the normally closed (N.C.) position but in series with any flame supervision device. *(See Flame Supervision Devices – 6.9).*

The signals created at the micro switch are transmitted to the printed circuit board usually by 3 wires but some combi boilers use 2.

3 wire – C = common N.O. = normally open N.C. = normally closed

2 wire – C = common N.C. = normally closed

Common wire (C) – Printed Circuit Board to Air Pressure Switch

Normally Open Wire (N.O.) – Air Pressure Switch to Printed Circuit Board once the fan is operating and negative pressure is established.
Normally Closed Wire (N.C.) – Rest or reset mode and must be proved at the printed circuit board before next ignition sequences commenced.

Fault Finding
If the fan begins to fail or its performance is impaired the differential pressure will reduce. This reduction in the differential pressure will be transmitted via the air pressure hoses to the air pressure switch. This will result in the air pressure switch resuming a 'no air' position (N.C. – normally closed) and the main gas valve closing, shutting down the appliance. Remember:-

No Fan = No air proving at Air Pressure Switch = No Ignition Sequence

Any fault will be at or before the fan. Always follow safe gas and electrical procedure.

1. The air pressure switch is prone to failure and when it fails the ignition sequence stops at 'fan running.' Listen/watch for ignition sparking. If sparking is present the fault is at or after ignition
2. If a fan assisted combi boiler will not start the ignition sequence, always check the fan is running before checking the air pressure switch
3. Check for blockages or partial blockages of the flue/airways
4. Check the air pressure switch 2 or 3 electrical connections are secure and terminals make good contact
5. Check the air pressure hoses are attached firmly and not kinked. They can perish and split and can get condensation inside. If a brass insert is fitted at the air pressure switch end of the hose ensure the small hole is not blocked
6. Check the air pressure hoses are attached to the correct +ve or –ve ports of the air pressure switch
7. Check for damaged/perforated diaphragm with air pressure switch. Use manometer on either port and it should read about ± 2mbars depending on which port is used
8. The electrical micro switch is invariably contained within the air pressure switch housing. If it fails the whole air pressure switch will need to be replaced. A faulty switch will be identified by the procedure below

9. Check the combustion chamber and appliance case seals are intact and in good condition. Replace if required
10. Check pilot window glass and seals are intact. Replace if required
11. Check the ignition lead terminals are attached correctly *(see Ignition/Pilot Systems – 6.14)*

To prove the fan is running effectively, the air pressure hoses are in good condition and attached correctly and the air pressure switch is functioning, electrically test signals at the printed circuit board using an approved test meter.

(a) Follow the common wire from the air pressure switch to the printed circuit board
(b) Measure between common wire (C) and normally closed (N.C.). This should read zero volts
(c) When there is a call for heat 'demand' this will become 230V (N.C.)
(d) N.O. will become 230V if signal from fan is proved
(e) It is advisable to check across neutral to C and N.C. and N.O
(f) If the fan, air pressure hoses, air pressure switch, terminals and leads are functioning correctly there will be 230V at the N.O. to neutral on the printed circuit board. Testing this measurement first will eliminate these components from the fault finding procedures

Before fault finding the air pressure switch always check:-

1. The combi boiler ignition is not sparking (listen/watch)
2. The fan is running at the correct speed
3. Flue/airways are not blocked or partially blocked

Fan Running O.K./No Ignition Sparking

Reason	Repair
Electrical connectors loose	Firmly attach terminals and ensure good contact – *6.11*
Air pressure hoses loose	Firmly fix air pressure hoses to ports – *6.1*
Air pressure hoses split/perished	Replace air pressure hoses – *6.1*
Condensation inside air pressure hoses	Remove, blow through or replace – *6.1*
Brass insert at port is blocked/partially blocked	Remove hose and blow through brass insert – *6.1*
Air pressure hoses +ve/-ve pressure connections reversed	Correctly re-position hoses – *6.1*
Air pressure switch diaphragm damaged/perforated	Replace Air Pressure Switch – *6.1*
Electrical micro-switch faulty	Replace Air Pressure Switch – *6.1*
Combustion chamber case seals in poor condition	Replace seals – *6.11*
Pilot window glass broken	Replace glass – *6.11*
Pilot window seals in poor condition	Replace glass seal – *6.11*

Servicing and Repair

The air pressure switch and the micro switch it contains require little or no maintenance. If it fails it is easily replaced:-

Air pressure switch and mounting bracket

1. Electrically isolate the combi boiler
2. Note the wire colour coding and positions of the 3 wires and remove. If these wires are incorrectly replaced the printed circuit board may be damaged
3. Note the connection points of the air pressure hoses and remove
4. The air pressure switch is usually secured by 2 screws holding the mounting bracket. Remove screws
5. Fit the new air pressure switch using the 2 screws to firmly attach the mounting bracket. DO NOT use the body holes of the air pressure switch for mounting purposes
6. Replace air pressure hoses
7. Replace 3 wires
8. Carry out safety check and turn on the combi boiler

Should the air pressure hoses be perished or perforated they can easily be replaced. Use old hose as a template and cut new hose to length. Use correct hose and cut squarely. Ensure correct connection to the +ve and –ve ports. Also ensure they are not loose on the port (must be airtight) or pressure can be lost and the whole ignition sequence will stop at 'fan running.'

Testing air pressure switch with multimeter probes

Introduction

Also known as an automatic air eliminator or automatic air release valve the automatic air vent is now incorporated into most combi boilers. The provision of a built in automatic air vent is adequate when combined with a manual or auto air vent(s) fitted at the highest point(s) in the system.

The automatic air vent installed within the combi boiler case is designed to vent:-

1. Air dissolved in fresh water introduced into the system
2. Air trapped in system e.g. at the top of radiators which is slowly absorbed into the system water
3. Hydrogen produced by electrolysis within central heating system *(see 'Central Heating – Fault Finding & Repair' book)*

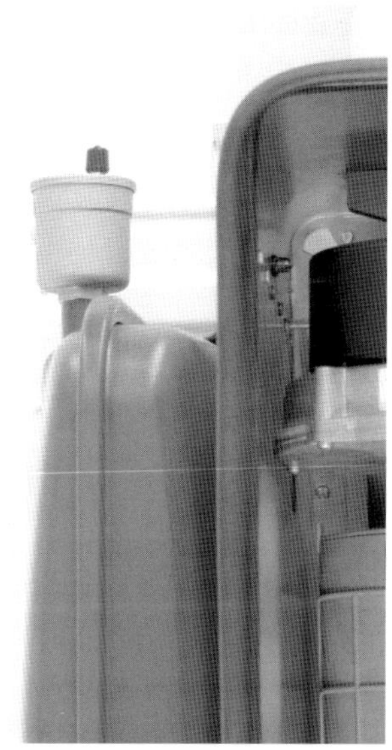

Automatic air vent on expansion vessel

Design and Contruction

The vent consists of the body, float, needle valve and cap and is manufactured to withstand a maximum pressure of around 6 bar and maximum temperature of 110°C approx.

Operation

Automatic air vent

The automatic air vent is designed to allow air to escape from the system by allowing a float to rise and fall with the water level in the system (or part of). Held up by the water the float forces a washer against the outlet seating. Air (or hydrogen) entering the valve will force the water level down so the float lowers. With the float lowering the washer is moved away from its seating and the air can now escape through the outlet. This allows water to enter the float chamber, forcing the float to rise and the washer to close off the outlet.

Fault Finding

The air vent is only designed to vent a small volume of air (or hydrogen). If the central heating system is not adequately vented at commissioning, following radiator removal or other work, trapped air may cause the combi boiler to shut down. This is due to air activating the low pressure switch, which senses a drop in pressure and to preserve the integrity of the main heat exchanger, sets the combi boiler to shut down *(see Low Water Pressure Switches – 6.15)*. Alternatively, air within the heat exchanger may cause the high limit thermostat to be activated, leading to shut-down.

Vent Cap Closed – This component will vent air automatically so the vent cap should be left open by several turns. It will not leak water when working correctly. Always check the vent cap has not been closed off by the householder (or engineer!) and if it has, vent the system and leave the vent cap open.

Seized Float and/or Needle Valve – Rust/sludge/scale can build up and harden causing the float to jam tight within the float chamber. This fault has the same symptoms as a closed vent cap. The component must be removed and cleaned or replaced.

Water Escaping Through Vent Cap – This will only occur when the valve is not closing correctly onto the valve seat. This can be due to:-

1. Rust/sludge/scale stuck to valve or valve seat preventing full closure
2. Rust/sludge/scale obstructing the free movement of the float within the float chamber. The component must be removed and cleaned or replaced

6.2 Automatic Air Vents

The escaping water may cause an electrical failure. The condensation caused within the casing may cause rusting or an electrical failure. There may, of course, be water damage to the property or fittings. This fault must be corrected quickly.

Combi Boiler Shutting Down Irregularly

Reason	Repair
Vent cap closed	Manually vent system and open vent cap several turns – *6.2*
Seized float/needle valve due to rust/sludge/scale	Remove and clean or replace vent – *6.2*

Water Escaping Through Vent Cap

Reason	Repair
Valve seating obstructed by rust/sludge/scale	Remove and clean or replace vent – *6.2*
Free movement of float obstructed by rust/sludge/scale	Remove and clean or replace vent – *6.2*

Servicing and Repair

Automatic air vent

When venting a central heating system always vent in this order:-

1. Radiators starting at the lowest level
2. Auto or manual air vent(s) at top of pipe work (usually in roof space)
3. Boiler heat exchanger and all other combi boiler vents as fitted
4. Pump – undo vent screw and turn impeller *(See Pumps – 6.18)*
5. Unscrew cap on the appliance auto air vent

There is usually an escape of water whenever venting is carried out so use a cloth for spillages and always protect and cover electrical leads, circuits, terminals and printed circuit boards.

At service check the vent cap is in the open position and not leaking water. If it is necessary to remove the air vent for cleaning or repairing a typical sequence might be:-

1. Drain the boiler/central heating system
2. Protect all electrical components within combi boiler
3. Remove circlip as fitted and unscrew the air vent from pipe
4. Clean or replace air vent ensuring thread sealant is used and/or 'O' ring and/or sealing washer as fitted is in good condition
5. Replace circlip as fitted
6. Re-fill and vent system at radiators and manual air vent(s)
7. Loosen automatic air vent cap

Introduction

Older central heating systems may have a fixed By-pass valve installed. Usually taking the form of a wheel operated gate valve connected across the flow and return pipes. Setting manual By-pass valves is a hit and miss affair unlike automatic By-pass valves.

It is now a requirement to fit an automatic By-pass valve, sometimes called a differential pressure valve, to all central heating systems unless the combi boiler manufacturers' instructions specifically say otherwise. It is positioned after the boiler and pump but before any 'tee' and the first radiator in the flow and as the last connection before the boiler in the return.

The main function of an automatic By-pass valve is to maintain a minimum flow rate through the heat exchanger. Minimum flow rates vary from model to model and heat output capacity with a condensing appliance of 24 kW output requiring approximately 775 litres per hour, a 30 kW 1050 litres per hour and a 38 kW 1200 litres per hour. A rule of thumb is 20 litres per minute minimum with non condensing combi boilers requiring less. This protects the heat exchanger and contributes to the combi boiler efficiency and helps prolong the life of the circulator *(see Pumps – 6.18)*. This leads to fewer breakdowns and efficiency is increased which reduces the heating costs. As minimum flow rates vary considerably always follow the manufacturers instructions.

Design and Construction

They are non-electric and are normally closed. Typically designed with a maximum working pressure of 10 bar and a maximum working temperature of 110°C. Automatic By-pass valves are robust, spring loaded, self acting water valves designed to be installed between the flow and return pipes as far away from the boiler as possible after the pump but before the first radiator or any 'tee' in the flow pipe and as the last connection in the return pipe.

Automatic By-pass valves

Available with 15, 22 or 28mm compression fittings the setting adjustment range is normally between 0.05 – 0.5 bar (1 to 5 metre head). For combi boilers rated up to 19 kW on the central heating side, a 15mm By-pass valve and connecting pipe is required. For combi boilers rated at more than 19 kW (central heating) a 22mm By-pass valve and connecting pipe should be used.

Some combi boilers using a modulating circulating pump MUST NOT have a By-pass fitted. Always follow the manufacturers' instructions.

Note:- Do not install an automatic By-pass valve within 200mm of a bend. Also, By-pass valves are not bi-directional so install with the correct flow direction as indicated on the valve body.

Operation

Central Heating systems will have varying heating loads as zone valves and/or thermostatic radiator valves open and close. Fitting a By-pass valve will ensure a re-circulation proportional to the number of valves that close avoiding noise, such as kettling and maintaining the head of the pump at a constant value.

The automatic By-pass valve works on the principle that pump pressure will increase when flow paths are reduced. So when zone valves and/or thermostatic radiator valves close, system circulation pressure increases. This increase in pressure forces the By-pass valve to open against the spring. This allows for a constant flow (above minimum flow rate) through the heat exchanger. Once set to the required pressure its operation will automatically ensure a constant flow through the boiler, maintain an even system pressure, reduces system noise such as kettling and water hammer and accommodate circulating pump over-run.

Thermostatic Radiator or Zone Valve closes – reduces flow rate – increased pump head – increases system pressure – By-pass valve opens – minimum flow rate maintained through heat exchanger.

6.3 Automatic By-pass Valves

.Fault Finding
When zone valves and thermostatic radiator valves are all open and there is a demand for heat the automatic By-pass valve should be closed. As the heat demand is satisfied the various zone valves and/or thermostatic radiators valves will close, the flow resistance will increase causing the valve to open against the spring. This allows the minimum flow through the combi boiler to be maintained.

Automatic By-pass valve set at almost/fully closed – The combi boiler may shut down or go to lockout due to 'overheat' of the water in the heat exchanger.

Automatic By-pass valve is set at almost/fully open – The last/highest radiator(s) may not reach temperature as the resistance of the By-pass must offer sufficient resistance to allow system water to flow through the last/highest radiator(s).

Combi Boiler Shuts Down/Lockout Before 'Demand' is Satisfied

Reason	Repair
Additional radiators with thermostatic radiator valves	Increase automatic By-pass valve setting – *6.3*
Additional thermostatic radiator valves added	Increase automatic By-pass valve setting – *6.3*
Last/highest radiator not reaching temperature	Increase automatic By-pass valve setting – *6.3*
No automatic By-pass valve fitted	Fit automatic By-pass valve and set correctly – *6.3*
Room/programmable thermostat set too low	Re-set room/programmable thermostat – *6.21*

Servicing and Repair
These valves are designed to be a 'fit and forget' component within the central heating system. To achieve the desired effect the correct setting is important and is simply achieved:-

(a) With system cold, set automatic By-pass valve to maximum setting
(b) Set controls to create demand for heat
(c) Switch on combi boiler
(d) Reduce setting until valve just open i.e. valve starts to get warm
(e) Turn adjuster back to just close valve
(f) Check last/highest radiator is to temperature
(g) Lock off adjustor at that setting

Introduction
Gas, oil and coal are hydrocarbons and when burnt they combine with oxygen in the air and form water vapour. In a conventional boiler this vapour leaves the boiler with the combustion gases via the flue and into the atmosphere.

When water is heated its temperature rises until it reaches 100°C when it starts to boil. Up to this point the heat applied is referred to as sensible heat. If more heat is applied the water remains at 100°C and steam is produced. The extra energy applied to produce the steam has to go somewhere and it is contained within the steam. The heat trapped in the steam is called 'latent heat' – from the Latin 'latent' or 'latere' meaning 'be hidden' or 'hidden.'

Latent heat = The heat required to convert a liquid to vapour.

When steam is converted back to water the energy it contains i.e. latent heat, must be given up. Many days a year this happens naturally. When warm moist air cools to below 55°C it condenses, a dew is formed and this is known as the dew point. The latent heat is released into the atmosphere.

Traditional Boilers
A traditional fan flued boiler emits flue gases at a temperature between 150–250°C. Expelled with the flue gases is the water vapour produced from the combustion of the fuel and with that is the latent heat the vapour contains. This lost heat means that the boiler can only have a maximum efficiency of approximately 85%.

Condensing Boilers
If flue gases are cooled to the dew point, 55°C, the water vapour will condense to a liquid the condensate. Also, the latent heat contained within the water vapour will be released.

This is achieved by passing the flue gases over a larger or additional heat exchanger, dropping the flue gas temperature to approximately 55°C and allowing the heat exchanger to capture some or all of the latent heat. This raises the efficiency to 90% or higher.

Condensate
By cooling the flue gases and water vapour to 55°C or below the latent heat can be used but the water vapour turns into a liquid – the condensate. Natural gas will produce about 1.6 litres of water per cubic metre. A 35 kW boiler could use about 4 cubic metres of gas per hour so theoretically could produce about 6 litres of condensate per hour. This does not usually occur as the boiler rarely has to work flat out for this duration.

Nevertheless, 2 or 3 litres per hour of condensate could be formed and it has to be removed from the boiler. Unfortunately, condensate is slightly acidic with a pH ranging from 3.6 – 5.0 similar to tomato juice, which can over time corrode many metals.

Condensate sump

Design and Construction
As the condensate is acidic the sump/siphon and condensate discharge pipe must be in plastic (PVC, PVC-U, ABS, etc) pipe and all joints should be solvent weld (glued). DO NOT use compression joints.

Operation
Older condensing boilers allow the condensate to trickle freely away to the discharge point but this method is prone to the condensate freezing with a potential ice plug forming. This can lead to boiler lockout so any external pipe work must be insulated.

Condensate Discharge System

Many modern condensing boilers use a siphonic action sump and condensate trap. The siphon releases a fixed amount of condensate, around 100ml, avoiding the constant dripping/trickle and freeze problems associated with older style discharge systems.

The sump or siphon is built into the boiler but the condensate pipe work run has to be considered and installed by the installer.

The drainage pipe should be at least 22mm diameter (over flow pipe size) and must be installed with a minimum 2.5 degree fall. That's about 45mm for every metre run. The options for condensate pipe run to discharge terminal can be:-

(a) Into external gully (maximum run is 3 metre)
(b) Into internal soil stack (no length restriction but using a 75mm minimum trap)
(c) Into sink waste before sink trap (no length restriction but using a 75mm minimum trap)
(d) Into sink waste after sink trap (no length restriction but using a 75mm minimum trap)

Where possible choose an internal run as avoiding external runs also avoids freeze problems. For any external runs it is good practice to step up the 22mm pipe to 30mm pipe ($1^1/_4$ inch). Insulate external pipe if there is a freeze risk.

Internal waste system

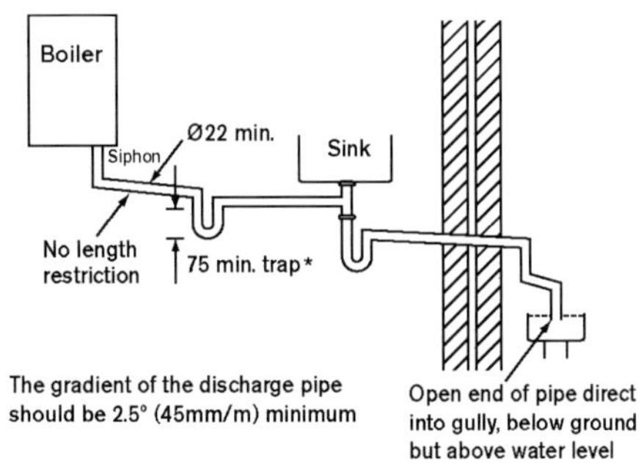

External waste system

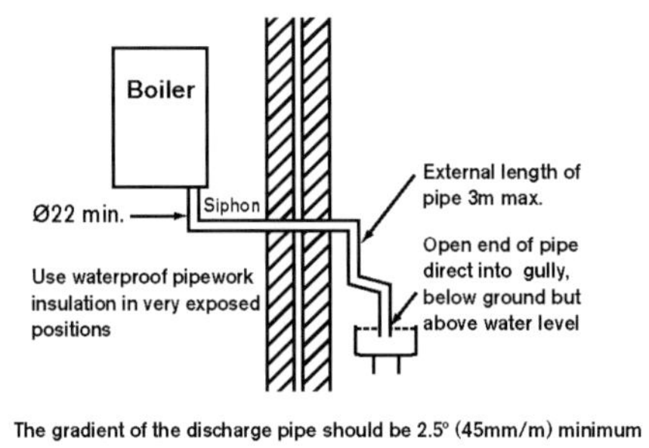

The rainwater system

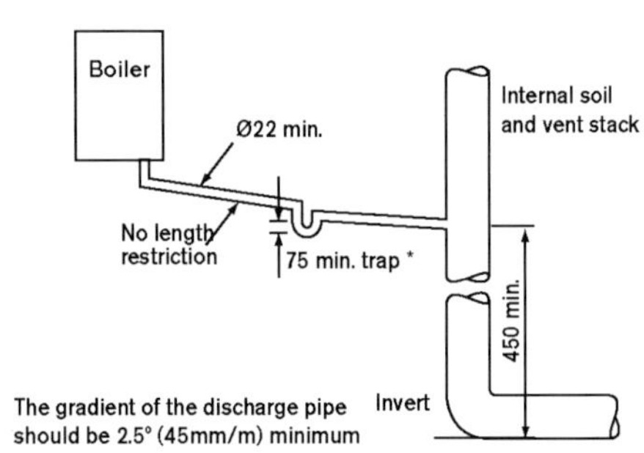

External purpose made soakaway

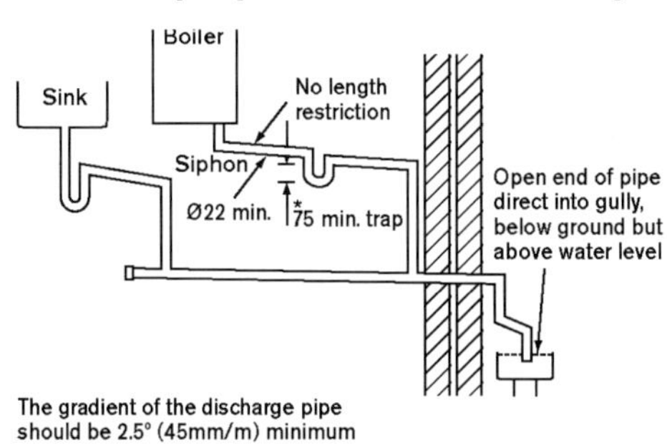

Before commissioning a new condensing boiler it is usually necessary to prime the siphon sump and fill the condensate trap with approximately $^1/_4$ litre of clean water. Remember to always follow the manufacturers instructions.

Fault Finding

Faults with the condensate discharge system normally show as the combi boiler going to lockout as the condensate fails to run away. This is either due to an ice plug forming in the pipe work due to freezing conditions or the angle of run away of pipe work is insufficient to allow drainage.

Boiler Lockout

Reason	Repair
Condensate formed ice plug in pipe work	Remove ice plug (use warm water), insulate pipe work – *6.4*
Condensate not running away	Adjust condensate pipe work to steeper incline – *6.4*

The condensate contains traces of ammonium, lead, cadmium, chromium, hydrocarbons, copper, nickel, mercury, zinc, tin, vanadium etc. so needs to discharged correctly.

Service and Repair

At service, condensate sumps and traps should be completely and carefully removed so as not to spill any condensate (acidic). Pour away condensate and rinse with clean water. Prime sumps and traps and refit.

Introduction

Not all combination boilers use a diverter valve but most do. There are different types of diverter valves utilised. They are an essential component used to divert the heated primary water within the boiler from heating the central heating system water to heating the domestic hot water.

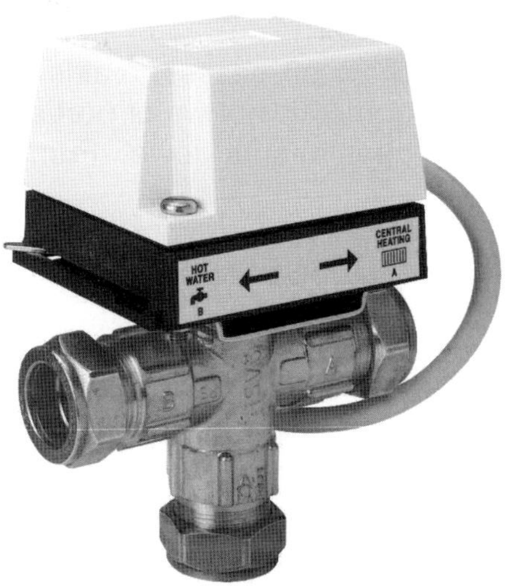

3 port diverter valve

The combi boiler will not heat central heating and domestic hot water at the same time as the diverter valve is either open or closed to each circuit. It is not designed to be a mid-position valve allowing a heat demand from both to be satisfied. A combi boiler will need and use its maximum heat output to heat the domestic hot water to the required levels. Some boilers may also modulate down relative to the flow at the draw off point. Gas burner modulation relative to the flow of domestic hot water or not, ALL of the combi boiler heat output capacity must be available whenever an instantaneous flow of domestic hot water is required. A combi boiler will not supply heat for a central heating demand if there is a demand for domestic hot water. It will always provide heat to meet a demand for domestic hot water before any call for heat from the central heating is met. It can be described as 'hot water' priority.

Operation

A combination diverter valve works in a similar way as a 230V 3 port fully diverted zone valve would work in a 'hot water' priority open/sealed traditional system:-

Open Vented or Sealed System Sequence

These systems have a mains supplied cold water storage cistern supplying water to a domestic hot water cylinder, a cylinder thermostat, 2 or 3 port zone valve(s), connecting pipe work, feed and expansion cistern for open vented or an expansion vessel filling loop and pressure relief valve for sealed system, pump and boiler. The operating sequence is:-

1. Hot water flow at draw off point supplied by domestic hot water cylinder
2. Hot water in domestic hot water cylinder cools as it is replaced by cold water from cold water storage cistern and the cylinder thermostat calls for heat
3. Cylinder thermostat signals to printed circuit board to commence firing boiler sequence
4. 3 port zone valve motors across, opening the domestic hot water port and closing the central heating port
5. Micro switches within the actuator activates pump via PCB
6. PCB continues sequence and boiler fires
7. The boiler heated water circulates through the heat exchanger within the domestic hot water cylinder
8. When cylinder thermostat is satisfied the pump and boiler are shut down
9. Valve closes off domestic hot water port and opens central heating port (valve in rest position)
10. Pump and/or fan may continue in over run

Combi boiler Sequence

3 port diverter valve

1. Hot water flow at draw off point supplied by rising main
2. Flow of water moves diverter valve, opening domestic hot water port and closing central heating circuit port and activates flow switch
3. Flow switch signals printed circuit board to activate pump and commence combi boiler firing sequence
4. Heated primary water heats mains supplied water flowing through domestic hot water heat exchanger (calorifier)

5. Domestic hot water thermostat controls domestic hot water temperature by controlling main burner
6. Draw off ceases, mains cold water flow stops, diverter valve moves to rest position closing domestic hot water port and opening central heating port. Micro switch de-activated and boiler goes to shut down. Pump and/or fan may continue in over run

Some combi boilers shut down for up to 30 seconds when moving from firing for central heating demand to domestic hot water or demand and vice versa.

Design and Construction
In early comb boilers a diverter valve similar to the familiar 3 port zone valve was used which comprised 2 parts, the 3 port valve body and the activating unit. Combi boilers diverter valves have evolved and there are 4 main types, including the modern 3 port valve, each comprising the valve body and an integral activating unit.

Diaphragm Pass Type Diverter Valve

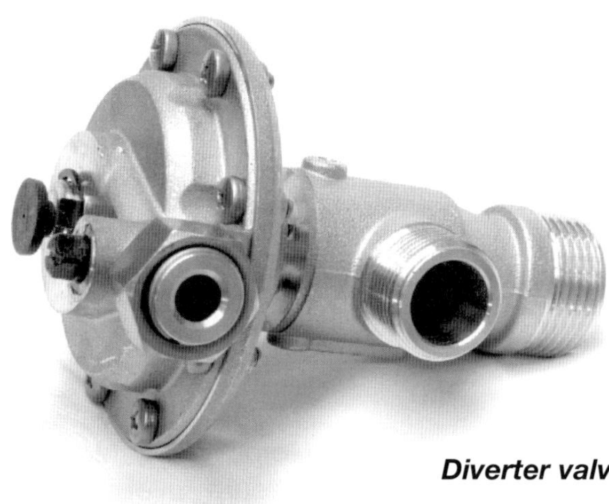

Diverter valve

Also known as a flexible diaphragm diverter valve (direct), this pressure operated diverter valve is the most commonly used. Consisting of a large diaphragm (typically about 70mm) and 5 ports heated primary water inlet, the central heating circuit outlet and the domestic hot water secondary heat exchanger outlet and cold water inlet and outlets.

When a hot draw off point is opened the drop in pressure created by the cold water at mains pressure passing through a 15mm diameter hole in the centre of the diaphragm causes the diaphragm to deflect upwards. This movement also causes the valve spindle to move upwards. The following happens:-

1. Lower valve closes off central heating port preventing the primary flow from the boiler entering the central heating circuit
2. Upper valve opens port to heat exchanger allowing the primary flow from the boiler to circulate through the secondary heat exchanger *(see Heat Exchangers – 6.13)*
3. Activates the micro switches on the head of the diverter valve to open and/or close sending a signal to the printed circuit board to commence ignition sequence

Note:- If a demand for domestic hot water is created when the boiler is firing in central heating mode some models shut down and re-fire whilst others remain 'fired.' The advantage with this valve is its simplicity and no electric motors. The most common fault with diaphragm pass diverter valves is the diaphragm splits. Fortunately, this can be replaced without draining down by removing the 8 nuts or 4 screws securing the diaphragm housing.

Venturi Type Diverter Valve
Also known as a differential pressure valve (indirect) this pressure operated diverter valve is probably the second most common. As the name implies the valve is based on the Venturi principle. This phenomenon was first described by Italian physicist G. B. Venturi who discovered that when a fluid is passed through a tube/duct which is wasp-waisted with flared ends pressure differentials are created. A high pressure is created at the Venturi inlet and a low pressure is created in the tube waist. This pressure difference can be used to operate the valve spindle.

The Venturi type diverter valve consists of a water tight chamber separated by a diaphragm. In the centre of the diaphragm are a bearing plate and a spindle (pin) which passes through a sealing unit. The diaphragm is designed to flex upwards when a pressure differential is created by water flowing through the Venturi, returning to its original position when the pressure equalises i.e. no water flowing.

As it flexes upwards. The bearing plate pushes the spindle through the sealing unit operating a micro-switch (flow switch).

As the valve spindle moves upwards the following happens:-

1. Lower valve closes off central heating port, preventing the primary flow from the boiler entering the central heating circuit
2. Upper valve opens port to heat exchanger allowing the primary flow from the boiler to circulate through the secondary heat exchanger. *(See Heat Exchangers – 6.13)*
3. Activates the micro switches on the head of the diverter valve to open and/or close sending a signal to the printed circuit board to commence ignition sequence

Diverter valve diaphragm

Again, a simple mode of operation with no electric motor is a benefit.
The most common fault is the diaphragm loses calibration or splits and the small Venturi tubes can become scaled or restricted by sludge/sediment/rust. In both faults the pressure differential is reduced and the diaphragm will not flex resulting in no domestic hot water.

Wax Capsule Type Diverter Valve
Also known as a thermal diverter valve this diverter valve is probably the third most commonly used. It uses the effects of heat on a volatile wax based fluid sealed within a brass capsule. A spindle is partly submerged in the wax. The capsule is positioned within the valve so when the central heating calls for heat the heated primary water flows, the capsule is heated, the wax expands and the spindle moves upwards. As it does it opens the primary port to the central heating and closes off the primary calorifier port. The spindle tip opens/closes a micro switch (flow switch).

When a hot water draw off point is opened mains cold water passes over the lower segment of the wax filled capsule. The water flowing over the capsule cools the wax which contracts. Under the action of a spring the spindle moves back into the capsule. This action closes off the primary port to the central heating and opens the primary calorifier port. Primary water flows to the domestic hot water secondary heat exchanger (calorifier) heating the mains fed cold water to set/flow temperature.

An advantage with the wax capsule type diverter valve is no electric motor to develop a fault but the wax capsule can loose calibration usually causing faults with the domestic hot water. This may show as sometimes hot sometimes cold depending on flow rate at the draw off point or the ambient temperature of the incoming cold supply etc. This fault only gets worse until eventually the boiler will not fire when there is a demand for domestic hot water.

Scale can build up on the capsule causing loss of sensitivity. This results in the wax capsule and spindle responding slowly to heat demand or not fully diverting, closing ports and operating any micro-switch (flow switch). Both of the above faults may cause the following fault:-

Domestic Hot Water Demand = Heating Circuit Warms

In both faults the wax capsule must be replaced which usually means the combi boiler must be drained down. As a temporary repair it may be possible to re-align the micro-switch.

Fault Finding
When tracing a combi boiler fault if the spindle moves out and in when the boiler is turned on and off and the micro-switch is made (listen for the clicking sound) the diverter valve is not faulty. If the spindle moves correctly and the micro-switch is not made, the switch arm is broken or the switch faulty. Replace the micro-switch on a like for like basis.

6.5 Diverter Valves

Ensure there is enough force in the spindle movement to make the micro-switch:-

1. Insert a screwdriver blade in between the spindle and micro switch. DO NOT bend the pin or damage the micro switch
2. Hold the screwdriver firmly
3. Turn the combi boiler on
4. As the spindle moves it must contact the screwdriver blade and firmly hold it against the micro switch. If the contact is timid suspect a fault
5. Turn the combi boiler off
6. Carefully remove screwdriver

Always check the domestic hot water flow rate at a draw off point. A clogged up mains supply filter/strainer or a badly scaled domestic hot water heat exchanger will restrict flow rate. Check a cold water draw off point supplied from the mains to measure mains supply flow and pressure.

With electrical 3 way valves you can expect in heating mode 230V AC between terminals 1 and 2 and 20–30V AC between terminals 2 and 3. In hot water mode expect 230V AC between terminals 2 and 3 and 20–30V AC between terminals 1 and 2.

No Domestic Hot Water

Diaphragm Pass Type

Reason	Repair
Split diaphragm	Replace diaphragm – *6.5*

Venturi Type

Reason	Repair
Split diaphragm	Replace diaphragm – *6.5*
Obstruction in Venturi i.e. scale, rust etc	Remove diverter valve/clean and blow through passage way – *6.5*

No Domestic Hot Water or Domestic Hot Water not at Temperature and Central Heating Circuit Warms

Wax Capsule Type

Reason	Repair
Wax lost calibration	Drain down boiler and replace capsule – *6.5*
Capsule scaled	Drain down boiler and replace capsule – *6.5*

Spindle Does Not Move

Reason	Repair
Pump failed	Repair/replace pump – *6.18*
Venturi pipe work blocked	Clean and blow through Venturi pipe work – *6.5*
Filter/strainer blocked	Clean strainer. Check water governor – *6.22*
Diaphragm perished/split	Replace diaphragm or diverter valve – *6.5*

Slow Movement of Spindle

Reason	Repair
Pump failing	Repair/replace pump – *6.18*
Spindle shaft scaled due to leak	Drain down boiler. Clean spindle or fit new spindle and sealing unit – *6.5*
Venturi partially blocked due to rust or sediment	Drain down boiler, dismantle valve, clean and blow through Venturi tube – *6.5*
Water flow slow due to sealed domestic hot water heat exchanger	Drain down boiler. Replace domestic hot water heat exchanger – *6.13*
Mains pressure low	Investigate reasons. Check water governor – *6.22*
Pressure loss at spindle sealing unit	Repair or replace unit or valve – *6.5*
Air may be trapped in valve after repair	Slacken unions to vent air, retighten – *6.5*

Spindle Moving but no Response

Reason	Repair
Not enough spindle force due to faulty diaphragm causing pressure loss	Drain boiler/replace diaphragm – *6.5*
Not enough spindle force due to sealing unit leaking	Drain boiler/replace sealing unit – *6.5*
Micro-switch faulty	Replace micro-switch – *6.5*
Printed Circuit Board not sequencing, fan not working, ignition sequence (sparking) or faulty gas valve etc	Prove micro-switch is not faulty and trace fault to next possible component – *6.6*

Service and Repair

At service check all unions for leaks or signs of a leak. Especially check the spindle where it passes through the sealing unit for a water leak. The water may dry on the spindle leaving salts and scale deposited on the spindle shaft. These deposits cause friction and slow the movement of the spindle leading to faults.

Introduction

Electronic boards are commonly referred to as a P.C.B. or printed circuit board, control board or control P.C.B. or as a sequence controller.

Printed circuit board

It is easy to see why these various names have developed as the P.C.B. is an electronic board in which the circuitry controls the sequence of events within the combi boiler. It receives input voltages and releases output voltages in a controlled safe repeatable pattern. This pattern is always the same unless a fault develops.

Design and Construction

There are as many variations of PCB's as there are combi boiler models so it is impossible to know exactly how each board works. What is needed is an understanding of how the board will sequence events. The PCB may be as 1 or 2 boards. Either way, the board(s) can be thought of in two parts:-

The thermostatic board sequencing live supply to the combi boiler when a demand for central heating via a central heating external controls closed circuit or a domestic hot water draw off demand is created. The ignition board sequencing live supply from thermostatic board with the ignition sequence.

On many PCB's there are potentiometers, also called 'pots' which have various applications e.g.

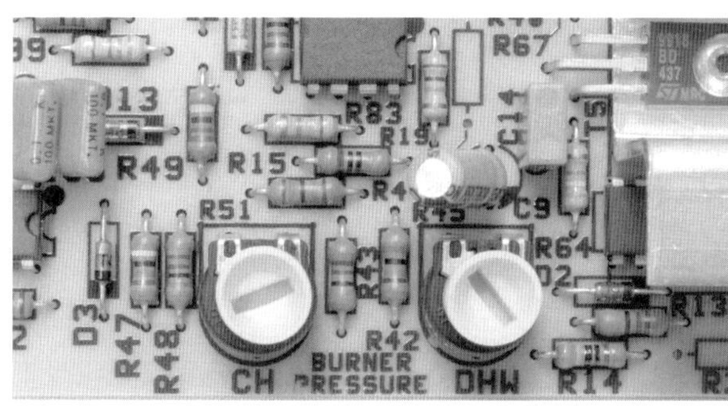

Central heating (CH) and domestic hot water (DHW) potentiometers

1. Range rating – setting maximum heat output to the size of the property
2. Ignition rate – adjusting the ignition speed
3. Low/high gas rate locking – by fixing the minimum or maximum fan speed the combi boiler will not modulate, and will cycle on the user controls set point similar to an on/off boiler. The potentiometers are factory set for the UK. DO NOT ADJUST unless the manufacturers' instructions specifically say otherwise. By adjusting these you may put the combi boiler out of specifications

Potentiometers

Dip switches also appear on many combi boilers. Usually as a block of three or four. They can be used to adjust pump use relative to the burner and also for setting a maximum heating point for applications such as under floor heating. Again, all dipswitches are factory set for the UK. DO NOT ADJUST unless the manufacturers' instructions specifically say otherwise. By adjusting these you may put the combi boiler out of specifications.

Operation

There are many sequence variations, component differences and voltages used but a typical sequence might be:-

230V to Appliance – Indicator 'on'

230V to PCB – PCB fuse is OK

Central Heating and Domestic Hot Water selected – indicator 'on' – call for central heating

Thermostatic Board Sequence

(a) PCB sends 230V to high limit stat – High limit stat returns 230V to PCB
(b) PCB sends 6V/230V to water low pressure switch (if fitted) – Water low pressure switch returns 4.5V/230V to PCB
(c) Provided the hot water flow switch (24V) is in open circuit (i.e. no demand for domestic hot water) – The diverter valve motors to open the central heating port (at rest) micro switch with its spindle tripping the micro switch, sending signal to PCB
(d) PCB sends 230V to pump – Pump runs
(e) Central heating flow switch (24V) senses flow and moves from open circuit to closed circuit – Sending signal back to PCB confirming water flow i.e. pump running
(f) PCB now sends signal – 230V to ignition side of PCB

All the above takes place in less than 1 second.

Ignition Board Sequence

(a) The ignition board or the ignition side of PCB receives signal providing 3 Amp fuse is OK (if fitted)
(b) PCB sends 230V to overheat thermostat – If overheat thermostat is OK then 230V returns to PCB
(c) PCB sends 230V to APS (common terminal) – APS sends ' no air' signal back to PCB if open
(d) PCB sends 36V dc/230V to fan – Fan starts creating differential pressure. This is sensed by flow switch or APS
(e) Air pressure switch or flow switch senses pressure changes – Sends 230V or low voltage back to PCB (proving air)
(f) PCB sends 22/24V to gas valve – Sparking commences
(g) Main burner lights – Spark ignition is suppressed and main gas valve opens
(h) Main burner flame is controlled by flame rectification device – Modureg modulates to central heating demand

If hot water draw off is opened, ignition side of PCB stays as is, some combi boilers close down this ignition board returning to normally open position but the thermostatic side recognises:-

(a) Hot water flow switch (secondary flow switch) closes – Sending 24V to PCB
(b) PCB sends 24V to hot water limit thermostat – If 'open' returns 24V to PCB
(c) Domestic hot water thermistor in range – Motorised valve motors from central heating position (normally rest) to domestic hot water position
(d) Gas valve adjusts to maximum gas rate i.e. voltage changes via potentiometer and thermistor (i.e. to operator setting on fascia) – Hot water is run off at draw off point

Printed circuit board

6.6 Electronic Boards

Fault Finding

The pattern of input voltages and output voltages at the PCB is controlled and repeatable. This pattern will not change and will always sequence voltages in turn. The sequence will stop at any point if a fault develops. The fault will always cause the combi boiler sequence to stop at that point. Therefore the PCB is the inbuilt fault finder. The fault will always be at the point it stopped at or immediately before. Using fault finding logic the faulty component or system fault will be quickly identified.

When a possible PCB fault is suspected check for small cracks/splits in the board circuit using a magnifying glass. Look for:-

1. Blown fuse, obvious but test continuity with multimeter *(see Multimeters – 5.4)*
2. Loose wires
3. Loose connections
4. Evidence of overheating i.e. burn marks
5. Water or condensation formed on board

Remember to clearly mark and write down connection position for any terminals/harness plugs removed from the board. Many of these will fit other terminals on the board!

The PCB is not repairable and if it does not supply the correct voltage will have to be replaced. However, do not remove the board unless you have proved it is faulty and is to be replaced.

Service and Repair

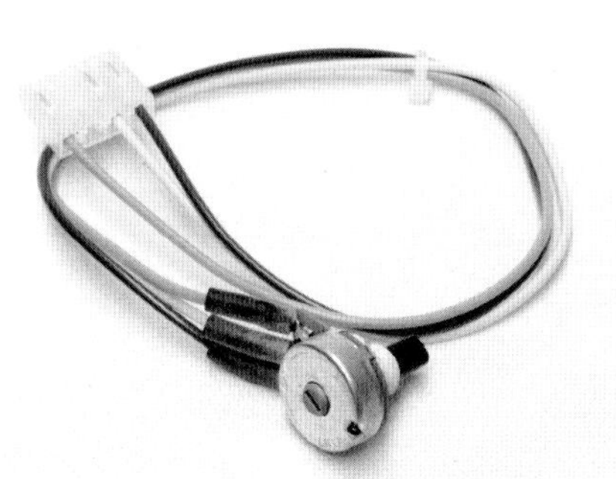

PCBs require no servicing. However, do not allow water to spill onto the board, terminals or wires. If the board becomes wet or is dirty DO NOT use a cloth or brush. Blow cold air over the board using a hairdryer! Most boards are not repairable and if faulty will have to be replaced. Replace with the exact board and make all connections as for the original board. Remember to mark ALL wires/connections and write down PCB connections. Carry out electrical safety check before firing the combi boiler.

Potentiometer

Introduction

In a traditional open vent system, i.e. open to the atmosphere, as the primary water is heated the increase in water volume is accommodated by the feed and expansion cistern. In a sealed system, i.e. closed to the atmosphere, the volume increase must be accommodated by an expansion vessel. Without it, system pressure will increase until 3 bar is reached, the pressure relief valve opens and system water is discharged. This pressure may also cause leaks to develop. As the water cools air will be sucked back into the system to replace the discharged water causing a list of faults.

To accommodate the expanded water a vessel is used which allows for a controlled increase in system pressure and also accommodates some of the expanded water. This protects the system from pressure surges and excessive operating pressures whilst maintaining a positive pressure within the system.

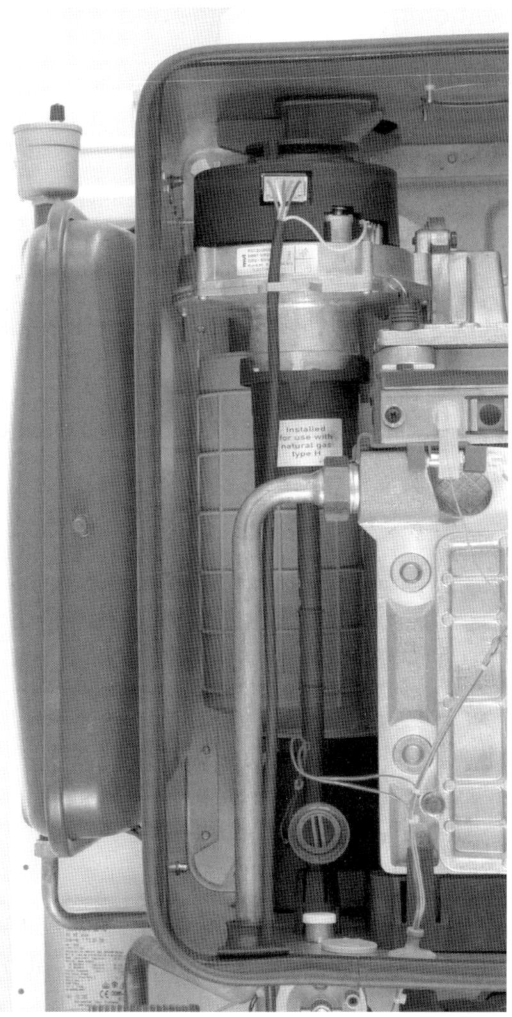

Expansion vessel side mounted for easy access

Nearly all modern combi boilers have the expansion vessel fitted within the casing. The manufacturers instructions will specify the maximum central heating system capacity for the vessel fitted, typically:-

8 litre vessel 90 litres central heating system capacity
10 litre vessel 130 litres central heating system capacity

If the volume of primary water contained within the central heating system is more than the maximum specified by the manufacturers an extra expansion vessel must be fitted. Failure to allow for additional primary water expansion will cause the following:-

1. Central heating system cold – Pressure at say 1 bar
2. Central heating 'call' for heat – Combi boiler fires
3. System warms – Pressure increases
4. System continues to warm – Pressure continues to increase to 3 bar causing pressure relief valve to discharge. Combi boiler continues to operate
5. Combi boiler may shut down and/or lockout
6. System cools – Pressure decreases
7. System cold – Pressure gauge reads zero and air will have been sucked into the central heating system to replace discharged water
8. Air in system will cause 'firing' problems due to low pressure switch etc. Also, this volume of air will cause rapid corrosion

If the system is re-pressurised using the filling loop and vented of all air this sequence of events will continue repeatedly until an extra expansion vessel is installed. Using the manufacturers maximum system capacity value and the actual capacity, the difference is the volume that must be accommodated when the system is heated. As water expands by approximately 4% when heated, 0–100°C it would seem correct to fit a vessel with 4% capacity of the over capacity volume but SEE NOTE BELOW.

If the pressure relief valve is set at 3 bar and the expansion vessel is charged at 1 bar the size of the expansion vessel can be calculated:-

Extra Water Content of System	Vessel Vol.	Extra Water Content of System	Vessel Vol.
25 litres	2.7	125 litres	13.6
50 litres	5.4	150 litres	16.3
75 litres	8.2	175 litres	19.1
100 litres	10.9	200 litres	21.8

Remember to size for the additional vessel, so only size for the additional water content:-

Actual capacity – Recommended maximum capacity = Additional water content

Note:- The quoted expansion vessel volume size is the overall capacity not the actual water capacity. Allowing for the heat expanded air within the expansion vessel etc. the capacity may be reduced to approximately 50%. Therefore, an extra 50 litres will require a 5 litre vessel.

With correct boiler sizing it will be unlikely, in a domestic property, to have more than 100 litres extra primary system water. As expansion vessels are available in commonly used sizes, using a 5, 8 or 10 litre will suit nearly all 'over capacity' installations.

To calculate the water content of a system use the following tables:-

Water Content of Copper Pipe

Pipe size	Water content per metre
8mm	0.036 litres
10mm	0.055 litres
15mm	0.145 litres
22mm	0.32 litres
28mm	0.539 litres

Domestic Hot Water Cylinder

Cylinder capacity	Water Content of coil
96 litres	1.8 litres
120 litres	2.2 litres
144 litres	2.7 litres
175 litres	3.2 litres
210 litres	3.9 litres

Water Content of Radiators

Radiator manufacturers usually specify the water content as 'section' or 'metre length.' Capacities will vary with manufacturer and model so, always refer to the manufacturers instruction/specifications but typically:-

Radiator Height	Single/Double	Water Content/section	Litre/Metre length
400mm	single	23 litres	2 litres
400mm	double	46 litres	4 litres
500mm	single	26 litres	2.5 litres
500mm	double	52 litres	5 litres
600mm	single	3 litres	3 litres
600mm	double	6 litres	6 litres

Water Content of Combi Boiler

A combi boiler will contain between 2 – 4 litres of water within the heat exchanger(s) and pipe work.

In a 4 bedroomed property with say 12 radiators a quick calculation might be:-

4 bedroom house with 12 radiators @ say 5 litres		=	60.0 litres
30 kW combi boiler water content (no domestic hot water)		=	2.5 litres
Pipe work run as	28mm say 10 metres	=	5.4 litres
	22 mm say 20 metres	=	6.4 litres
	15 mm say 50 metres	=	7.2 litres
	Total	=	81.5 litres

It can be seen that if the output of the combi boiler is correctly sized to the property the fitted expansion vessel will be sufficient.

If an extra expansion vessel is to be fitted, install on the return side as close to the boiler as possible and on the negative side of the pump. The connecting pipe work to the vessel must be 15mm minimum with no restriction installed such as an isolating valve. DO NOT stress the pipe work. Use a bracket to support the expansion vessel.

Domestic Hot Water Expansion Vessel
On many older combi boilers a small expansion vessel, charged to about 3 bar, is installed. This will absorb pressure surges in the hot water and cold supply pipe when a hot water draw off point is closed. This prevents/reduces water hammer and pipe expansion noises. When a check valve is used close to the combi boiler cold water inlet it may be necessary to install this small expansion vessel in the pipe work.

Mini expansion vessel

Design and Construction
The expansion vessel is a carbon steel container which is internally divided into 2 sections by a flexible neoprene/SBR rubber diaphragm. There is usually a charging valve, long or short, fitted. This is a schraeder valve as fitted to car tyres. The vessel is factory charged with nitrogen on one side of the diaphragm to between .5 and 1 bar depending on make, size and specification. Air is used to re-charge the vessel on site, usually with a car foot pump.

If not fitted within the combi boiler the expansion vessel must be sized correctly, installed in the return pipe as close to the combi boiler as possible and on the negative side i.e. the inlet or suction side of the pump:-

Sized Correctly – To prevent system overheat/pressure increase and causing the pressure relief valve to discharge.

Installed In Return Pipe – The lower water temperature will help prolong the working life of the neoprene diaphragm.

Close To Combi Boiler – Return water at lowest temperature.

On Negative Side Of Pump – This avoids the action of positive pump pressure on the diaphragm.

The vessel must be well supported and connecting pipe work must be 15mm minimum with union to accept the usual 3/4 inch BSP male connection on the expansion vessel. DO NOT stress pipe work.

Operation
When the primary water is heated it expands by approximately 4% and the increased water volume is accommodated by the expansion vessel. The water in one vessel section presses against the diaphragm and as the pressure/volume increases the diaphragm flexes toward the other section compressing the nitrogen. As the water cools the pressure/volume decreases and the diaphragm returns to its original position, forcing water out of the expansion vessel back into the primary system.

The system pressure is measured using a pressure gauge, which is normally built in to the combi boiler *(see Pressure Gauges – 6.17)*. Use a car tyre pressure gauge to check the pressure of the expansion vessel.

6.7 Expansion Vessels

Fault Finding

When tracing a fault always check the pressure and temperature gauge. If the pressure relief valve is discharging the gauge will help with fault finding:-

Temperature gauge	Pressure gauge	
Normal	High	Fault is with expansion vessel/valve/ connecting pipe/domestic hot water heat exchanger/filling loop
High	High	Fault is with limit thermostat/pump/ scaling

Regular Topping Up of System

Reason	Repair
Excessive venting of radiators by householder	Establish reason for air in system, trace reason, correct and re-pressurise system – see 'Central Heating – Fault Finding & Repair' book
Leak in system at pump or radiator unions, pipe work, heat exchanger, etc	Raise system pressure to 3 bar using the filling loop, trace and repair leak, release water until pressure is at 1 bar – *6.17*

Regular Topping Up of System and Pressure Relief Valve Discharges
Pressure Gauge Low/Zero When Cold

Reason	Repair
Expansion vessel lost charge naturally over time	Drain down, re-pressurise vessel and top up system – *6.7*
Expansion vessel diaphragm split or porous	Check for water at schraeder valve when depressed. Replace expansion vessel – *6.7*
Schraeder valve faulty and letting out nitrogen/air	Check for air leak at valve with LDF. Re-tighten/replace valve if faulty – *6.7*
Pressure relief valve faulty/leaking due to obstruction at seating	Clear obstruction or replace pressure relief valve. Raise to 3 bar, operate pressure relief valve and reduce to 1 bar – *6.7*
Pressure over 3 bar, water discharge at pressure relief valve – expansion vessel undersized	Replace with larger or fit additional vessel – *6.7*
Vessel to pipe work connecting pipe blocked with sludge	Remove vessel and clear blockage or replace connecting pipe. Refit vessel – *6.7*
Pressure gauge reading low. Pressure relief valve discharges	Check/replace faulty pressure gauge and sensing head – *6.17*
Excessive fill pressure	Reduce pressure to 1 bar (cold) by releasing water from system – *6.8*

Regular Topping Up of System/Pressure Relief Valve Discharge/System Overheating

Reason	Repair
Combi boiler overheating due to scaling or sludge with pressure relief valve discharging	Drain, flush and protect central heating system – *2.5*
Combi boiler overheating due to pump failing/failed with pressure relief valve discharging	Repair/replace pump/pump head – *6.18*
Limit thermostat faulty so overheating to overheat thermostat and pressure relief valve discharging	Check/replace limit thermostat – *6.21*
Pump or fan over run faulty	Check pump – 6.18 Check fan – *6.11*
Manual or Automatic By pass set incorrectly	Reset/replace By pass valve – *6.3*

System Cold, Pressure Relief Valve Discharges and High Pressure at Gauge (3bar)

Reason	Repair
Domestic hot water heat exchanger leaking across to primary system	Replace domestic hot water heat exchanger – *6.13*
Filling loop attached and isolators letting by	Fully tighten isolators, remove filling loop and inform customer – *6.8*

Service and Repair

Charging expansion vessel:-

1. Isolate the combi boiler and drain down using isolating valves on flow and return *(see Pipe Freezers – 5.5)*
2. Locate schraeder valve
3. Remove cap
4. Use a car foot pump with pressure gauge to charge vessel to 1.2 bar
5. Depress valve briefly and check no water escapes indicating a perished/perforated diaphragm. If water does escape, replace vessel
6. Check for an air leak at the valve using LDF
7. Reduce pressure to 1 bar or manufacturers instructions
8. Replace cap
9. Open isolating valves, vent pump and recharge the primary system to 1.0 bar, using the filling loop, venting as necessary

Replacing expansion vessel:-

If the expansion vessel is suspected of being faulty with a possible split or porous diaphragm, depress the schraeder valve slightly. If water escapes/bubbles through valve the vessel will have to be replaced. Remember, if no water escapes during this procedure you have just de-pressurised the expansion vessel!

When the schraeder valve is suspected of leaking, using LDF will show air loss through valve. Either tighten or replace the valve. With many expansion vessels the schraeder valve is replaceable, after draining down. If the valve cannot be removed, replace the expansion vessel. ALWAYS check, using the car tyre pressure gauge, the full pressure of new expansion vessels is around 1 bar before fitting to the combi boiler. Adjust pressure as necessary.

6.7 Expansion Vessels

Replacing an integral expansion vessel may involve removing the combi boiler off the wall. In other combi boilers, components such as the air pressure switch, primary heat exchanger, domestic hot water heat exchanger and pump may all have to be removed giving access to remove the expansion vessel. In many modern combi boilers the vessel is removed by loosening a screw(s) and undoing the union connection. Do not loose the sealing washer if fitted.

In each case the combi boiler must be isolated and drained down *(see Pipe Freezers – 5.5)*. Always replace like for like. DO NOT undersize the expansion vessel.

Introduction

Conventional or open vent systems use a feed and expansion cistern to fill the central heating system with water and maintain water levels.

Filling link and key

A sealed system eliminates the requirement of a feed and expansion cistern. In the absence of a feed and expansion cistern a commissioning fill up point and a point to manually 'top-up' the central heating system from time to time is necessary. This is provided by the 'filling loop' sometimes called a 'temporary filling hose.' Some combi boilers use a large plastic 'key' to fill/top up the system. It fulfils the same function as the filling loop.

The fill point with stop valve is always tee'd on the central heating return pipe close to the combi boiler, usually directly below. A mains pressure cold supply pipe will run to the combi boiler to provide the source of potential hot water and will have a tee off of it and stop valve fitted. The filling loop is connected across the isolator unions from the mains supply and the central heating return.

To prevent contamination of the cold supply with the primary central heating system water a double check valve must be installed. On systems installed before December 2002 many of the double check valves were fitted on the central heating return side of the filling loop. From that date a double check valve MUST be fitted on the mains cold water side of the filling loop and a service valve must be fitted on the central heating return and cold water sides.

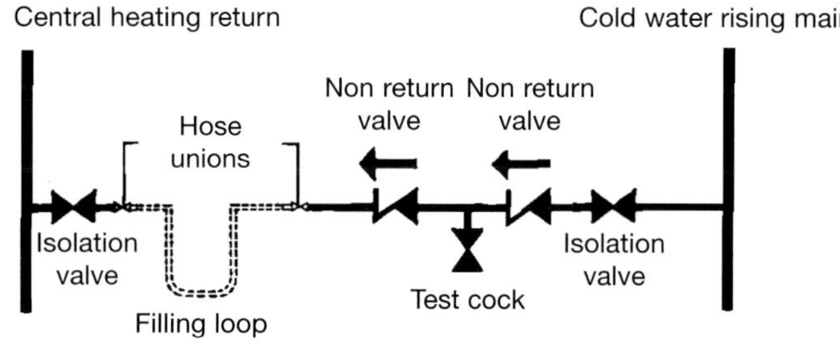

Filling loop showing isolation valves and non-return valves

It is a regulation that filling loops are disconnected after use to reduce the possibility of contamination. Suggest to the householder a sensible place to keep the filling loop so it does not get lost. An approved method of filling and top up must be used to prevent any mixing of potable water with the primary water in the central heating system. Always use a filling loop that complies with regulations.

A few systems do not use a filling loop. Instead a make up vessel is used. Located well above the highest point of the system it is filled manually or may be connected to the mains supply.

Design and Construction

Filling loops are usually metal braid type although, and some are plastic, all have 15mm connectors. Comprising a double check valve connected to either a straight or tee isolating valve. Some have an isolating valve at the double check end.

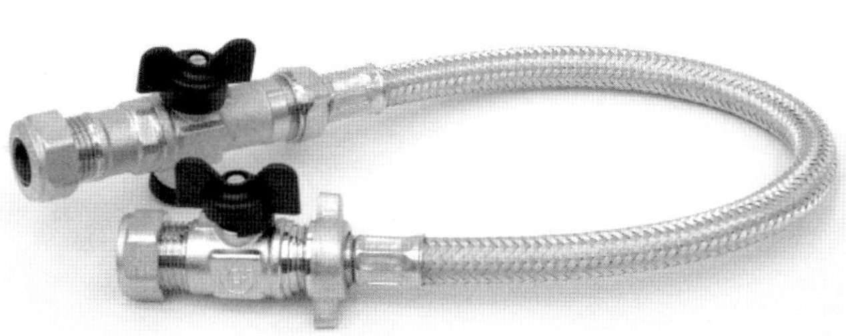

Filling loop

There are also available sealed system filling kits which contain all the essential components required for system installation. Containing a filling loop, pressure relief valve, pressure gauge, double check valve, isolators and an expansion vessel mounting bracket.

Operation

Used at commissioning and topping up of systems the filling loop is attached to the tees' on the cold supply and central heating return. Both isolators are opened and the pressure gauge observed until 1 bar or the recommended system pressure is achieved. Both isolators are closed and the filling loop removed, sealing open ends.

6.8 Filling Loops

Fault Finding
If the filling loop is always removed when not in immediate use, it will be impossible for faults to be due to this system component accessory. However, many filling loops are not removed by the householder between top-up fills. This may be because:-

1. The house holder has never been informed of the importance of removal
2. The importance of removal was not emphasised by the installing engineer
3. A 'loss of water' fault necessitates regular topping up and it is too much trouble to frequently attach/detach filling loop

Leaving the filling loop attached between 'fills' may cause faults to develop if the isolators are not fully closed or are letting by. Water at mains pressure will enter the system creating a higher than required system pressure *(see Pressure Gauges – 6.17)*. This may also be accompanied by the occasional discharge from the pressure relief valve, especially on heating. Fully close isolators, disconnect the filling loop and reduce the pressure to about 1 bar when cold.

Pressure Gauge Reads High

Reason	Repair
Attached filling loop letting by	Fully tighten isolators and remove filling loop – *6.8*

Service and Repair
At combi boiler service the filling loop may be used to 'top up' the system to the recommended pressure. Always remove the filling loop and inform householder of the importance of removal i.e. contamination and legal requirement.

The filling loop is robust and usually does not develop faults.

Introduction

A flame supervision device or FSD is a safety device used to prevent gas flow to burners when a fault occurs e.g. no pilot/main flame present. It will also ensure the safe ignition of a burner by preventing un-ignited gas from building up. It senses the presence of a flame when the gas valve is open.

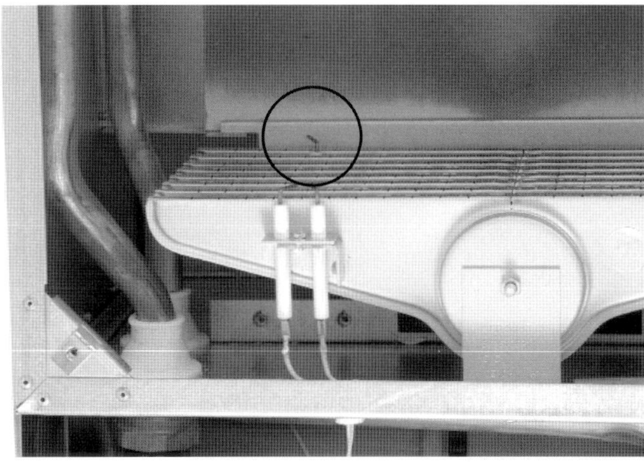

FSD set on burner

As a safety device it must have measurable criteria. This criteria is the maximum time the flame safety device will fail safe i.e. shut down the gas valve. This fail safe period varies with appliances and not with the device used. Flame supervision devices within combination boilers as with other central heating boilers and multi-point water heaters must:-

FAIL SAFE IN 60 SECONDS.

Design and Construction

There are several ways of detecting the presence of a flame including:-

(a) Using the heat from the flame:-
Bimetallic strips e.g. instantaneous water heaters
Liquid vapour type e.g. cooker ovens
Thermo-electric type e.g. some combination boilers

(b) Using the flame as a conductor:-
Flame conductance e.g. some boilers
Flame rectification e.g. many combination boilers

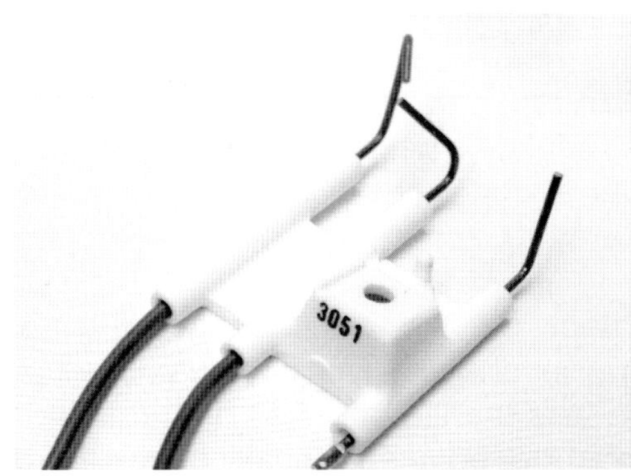

Flame supervision device

Thermo-Electric Type

Generally called a thermocouple it uses the principle of increasing molecular activity when heat is applied. When two dissimilar metals, usually iron and eureka, are joined together at one point and this point is heated, the different molecular activity within the two metals creates a small electric current. The two dissimilar metals are contained in a thin copper tube and the jointed tip positioned within the flame. The other end is connected to an electromagnetic valve. The electric current energises the electromagnet, holding the valve off its seating *(see Solenoids – 6.19)*.

Operation

The electric output (10–30mV) of the thermocouple is not enough to lift the valve off its seating. The valve must be opened manually and for enough time (10–20 seconds) to allow the pilot to light and to heat the thermocouple sufficiently to energise the electromagnetic valve, holding it off its seating. When the flame is extinguished the current will stop, the spring loaded valve will close, shutting off the gas to the main burner.

To keep the valve energised the electric current must flow uninterrupted. To maintain the current the thermocouple tip must be constantly heated, usually by the pilot flame. In fan assisted combi boilers the fan must run at low speed when the pilot is alight to clear the products of combustion produced by the pilot flame *(see Flue Fans – 6.11)* avoiding corrosion and possible flue gas spillage. In the absence of a flame the device will fail safe and the boiler will shut down or go to lockout.

6.9 Flame Supervision Devices

Fault Finding

Gas Valve not Energising

Reason	Repair
Tip not correctly positioned in flame	Check pilot flame is correct and/or reposition tip – *6.14*
Pilot flame too short	Clean pilot assembly – *6.14*
Pilot flame misaligned	Clean and/or realign – *6.14*
Tip burnt out	Replace thermocouple – *6.9*
Build up of soot/dirt on tip	Carefully clean tip – *6.9*
Insulation damaged causing short	Check resistance, high or infinity, faulty, replace thermocouple – *6.9*
Connection to gas valve loose	Remake connection – *6.12*
Electro-magnetic valve faulty	Check gas valve – *6.12*

Service and Repair

At service check the thermocouple tip is sufficiently heated by correct positioning within the pilot flame, check the pilot flame is not too short or misaligned. Further check the tip is not burnt away. Carefully remove any dirt or soot.

Check for physical damage causing a breakdown of insulation between the inner core and outer copper sheath. This can be caused by bending a radius too tightly. Check the resistance which should be less than 1 Ohm. A high reading or infinity proves the lead is broken causing an open circuit and must be replaced. Finally check the connection at the gas valve is well made and secure. The electric current produced can be measured with a multimeter set on milli-volts to measure the 10mV – 30mV output. Attach multimeter leads to inner and outer conductors at the gas valve connection. Hold in flame failure button and with pilot flame alight read output. A very low or zero reading proves a faulty thermocouple.

If the thermocouple is faulty it cannot be repaired and must be replaced, when replacing check the rating and fittings are the same.

Flame Rectification Type

Combustion is a chemical reaction and produces electrically charged particles which allow an electric current to flow through a flame.

Flame rectification devices use this ionised zone to rectify an alternating current (a.c.) to a direct current (d.c.) which, via a relay, signals the printed circuit board to open the gas valve.

Should the flame fail, the ionised zone of the flame will immediately be lost and the rectified electric current will stop flowing. This results in no signal to the PCB causing the gas valve to close. This shut down is usually within 1 second.

Operation

Positive ions are continuously formed in the pilot flame by part rectifying an a.c. current to d.c. current. This is detected by the printed circuit board energising the gas valve. Should the flame extinguish the gas valve will immediately close.

Fault Finding

Gas Valve not Energising

Reason	Repair
Probe not correctly positioned	Check position – *6.9*
Rectification lead faulty	Check continuity and replace if faulty – *6.9*
Faulty connections	Clean and remake connections – *6.9*
Dirt blocking burner close to rectification probe	Clean burner of fluff and dirt – *6.9*
Reversed Polarity	Check electric supply – *3.4*
Electromagnetic valve faulty	Check gas valve – *6.12*

Service and Repair

Some combi boilers using flame rectification as the flame supervision device use either 2 electrodes or an electrode and the burner pilot assembly. Ensure the probe is clean and in good condition and is correctly positioned within the flame for maximum current to flow. This current is measured in micro-amps and is not measured on site. The probe should be correctly positioned above the main burner or pilot with a 3mm gap.

Check the lead is clean and in good condition and check the connections are clean and sound. Clean the burner ports close to the rectification probe.

Testing a Thermo-Electric Flame Supervision Device

There are several ways to test a flame supervision device including 'click', 'blip' and 'mini tightness' methods. The click method simply entails firing the combi and then closing the gas isolation valve, listening for the click as the electro-magnetic valve shuts (within 60 seconds). Unfortunately, this method does not test for the valve letting by gas if it fails to close properly. Therefore, use either the 'blip' or 'mini tightness' methods.

Blip Method

1. Connect manometer to burner pressure test point
2. Light the combi-boiler
3. Turn off the gas supply at the isolation valve
4. Listen for the 'click' as the elector-magnetic valve shuts. This must be within 60 seconds or the flame supervision device is faulty – *see below*
5. Turn on the gas supply at the isolation valve
6. Observe manometer:-

   No movement (no increase in pressure) – Flame supervision device working correctly

   Some movement (some increase in pressure) – Flame supervision device allowing valve to pass gas and is faulty – *see below*

N.B. When a flame supervision device is faulty the combi boiler should be classified as immediately dangerous. Repair or isolate and make safe.

6.9 Flame Supervision Devices

Mini-Tightness Method

1. Connect manometer to inlet test point
2. Light the combi boiler
3. Turn off the gas supply at the isolation valve
4. Listen for the 'click' as the electro-magnetic valve shots. This must be within 60 seconds or the flame supervision device is faulty – *see below*
5. Turn on the gas supply at the isolation valve and raise the pressure to 20 mbar
6. Turn off the gas supply at the isolation valve
7. Observe manometer:-

   No drop in pressure – Flame supervision device working correctly

   Drop in pressure – Flame supervision device allowing valve to pass gas and is faulty – *see below*

N.B. When a flame supervision device is faulty the combi boiler should be classified as immediately dangerous. Repair or isolate and make safe.

Testing a Flame Conduction/Rectification Flame Supervision Device

1. Light the combi boiler
2. Turn off the gas supply at the isolation valve
3. The flame extinguishes and the combi boiler should spark at the electrode
4. Turn on the gas supply at the isolation valve
5. The gas should re-ignite

Note:- Some combi boilers will go to lockout if the flame is not re-ignited. If this is the case then re-set.

 Flow Switches 6.10

Introduction

Any component that responds to a flow and operates a switch could be termed a flow switch. This can cause some confusion and must be taken in context. There are many types of flow including:-

Primary water flow – Central Heating System

Secondary Water Flow – Domestic Hot Water

Air Flow – Combustion Air

Flue Gas Flow – Products of Combustion

The flow of primary/secondary water and air/gas/flue gases can be used to operate flow switches.

Operation

Widely used flow switch

The flow of primary water through the heat exchanger must be controlled. Insufficient flow will cause faults to develop and may damage the heat exchanger. A flow switch or micro-switch can detect water flow and signal a printed circuit board. The flow of secondary water can be used to operate a micro-switch or flow switch to signal the printed circuit board to sequence the boiler to maximum output. Air flow and flue gas flow can be sensed by the air pressure switch operating a micro-switch or by thermistors *(see Thermistors – 6.20).*

Diverter valves often operate a micro-switch. Many combi boilers have a double micro-switch mounted in a plastic housing. Mounted onto the diaphragm chamber of the diverter valve the micro-switch signals the printed circuit board with the operating status of the diverter valve *(see Diverter Valves – 6.5).*

Flow switch

When the valve is in the 'not activated' position the micro-switch is also not activated. A demand for domestic hot water or central heating will activate the valve, the spindle will extend out of the valve body and the spindle end presses the contacts on the micro-switches.

The signal to the printed circuit board may cause the pump, fan, gas valve and burner to switch on, if not already operating.

Flow switch

The flow of domestic hot water can be sensed by a mechanical water flow sensor. A turbine, supporting a magnet, rotates with water flow in front of a Hall effect sensor. The Hall effect sensor picks up the field of the magnet giving a pulsed output. The output is directly proportional to the water flow. Therefore, the printed circuit board can be signalled to modulate the burner load or to simply fire the burner when the flow of domestic hot water reaches the minimum flow rate.

Another flow switch

Constructed from temperature and pressure resistant plastic all parts in contact with water are made of the appropriate materials to conform to the regulations for potable water. A typical combi boiler water flow sensor is designed to operate between -20–85°C ambient temperature, with a minimum operating flow rate of 1–1.5 litres/minute and a maximum operating flow rate of 30–35 litres/minute, with an accuracy of ± 10%. These water flow sensors are tested to over 15 bar.

Water flow senors are robust components and are designed and tested for a minimum life of 100,000 on/off cycles (5 seconds On & 5 seconds Off) at 10 litres/minute at 45°C.
That's on and off about 20 times a day, every day, for 15 years.

6.10 Flow Switches

Fault Finding
Micro-switches operate from a very small movement and it may not be obvious if it is switching. Closely inspect that there is adequate provision for the micro-switch to operate e.g. sufficient spindle movement. If the switch is signalling check its resistance and check connections are secure.

Switch Not Operating

Reason	Repair
Switch contact faulty	Replace switch
Switch mounting loose	Secure mounting, ensure correct position
Switch incorrectly aligned	Re-align switch with spindle/pin. Ensure correct position

Small particles of sand or rust from the rising mains cold water supply may restrict free movement of the turbine/wheel. A mesh filter/strainer is often installed to prevent ingress of particles.

Service and Repair
Micro-switches are not repairable and if faulty must be replaced. To check, set multimeter to ohms and test across common and normally open and normally closed:-

Overload/infinity = Switch open

1 Ω or less = Switch closed

When the switch is closed:-

1 Ω–75 Ω = Dirt/debris on contact or switch pitted. Replace switch

Mega ohms/overload/infinity = Switch broken. Replace switch

If water flow sensors are replaced, remember to isolate the boiler, drain down or use a pipe freezer and replace like for like.

Introduction

Many combi boilers now incorporate a fan to assist with the movement of combustion gases through balanced flues. The increased velocity of flue gases means smaller diameter (cross sectional area) flues can be used and the flue runs can be horizontal and longer. This enables a greater variety of flue termination sites to be available and allows for sites to be more remote from the combi boiler.

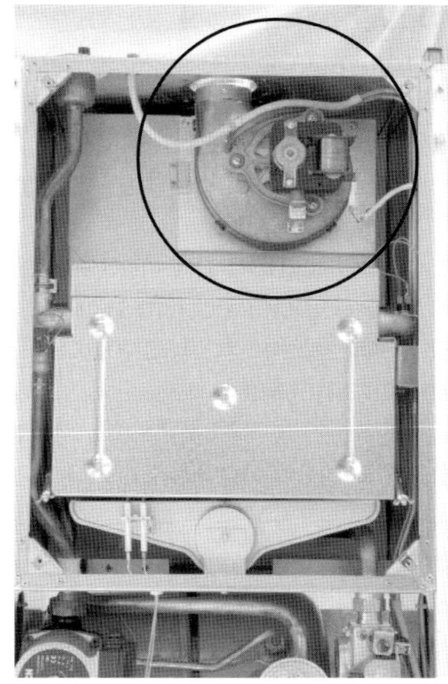

Flue fan

The flue gases are only removed if the fan is operating efficiently/correctly. Therefore, if the fan does not 'run' the combi-boilers control and safety device, air pressure switch or thermistors, will prevent the ignition sequence from continuing. Sparking will not commence and the gas valve will remain closed on its seating.

Design and Construction

The fan utilises a 230Va.c. (or 230Vd.c.) motor with centrifugal fan. On modern combi boilers the flue fan motor has sealed bearings which do not require lubrication. It is designed to work with high temperature flue gases. A standard efficiency combi boiler has flue gases at temperatures between 150–240°C. Condensing combi boilers flue gases are often at less than 55°C.

Fan speeds vary from model to model with typical minimum and maximum speeds of 1000 and 6000 rpm. The power of the motor is matched with the impellor within the scroll casing to cause the correct flow of flue gases to leave the boiler. The motor is often attached to a bracket that is mounted onto the steel scroll case via rubber anti-vibration feet. The inlet hole of the scroll coincides with the hole in the top of the combustion chamber flue hood. The fan outlet is secured to the boiler flue connection, often using a flexible resilient seal.

Note:- The fan design and layout is different for forced draught burners found in some condensing combi boilers.

Operation

When there is a demand for heat ('call') the fan will run. On older combi boilers with a permanent pilot it is usual that the fan runs at low speed to clear the products of combustion produced by the pilot flame *(see Air Pressure Switches – 6.1)*.

With the fan running, flue gases are drawn in through the centre ('eye') of a circular veined wheel (impellor). As the flue gases are thrown out of the fan a pressure drop is created which causes more flue gases to be drawn in. The faster the impellor spins the greater the centrifugal force thus creating a greater pressure differential. If the fan is turning at speed the differential pressure created will be transmitted via the air pressure hoses to the air pressure switch.

Alternatively the flue way thermistors *(see Thermistors – 6.20)* will sense a temperature change within the air flow and a signal will be transmitted to the printed circuit board – ignition etc.

Fan Over Run

When the demand for heat is satisfied many combi boilers use 'pump over run' to dissipate heat from the heat exchanger within the combustion chamber. This can cause problems with the residual heat being 'dumped' into the central heating systems. When the central heating is turned off this 'heat dump' can cause radiator(s) to become warm *(see Pumps – 6.18)*. Other combi boilers use a fan over run, forcing cool air over the heat exchanger preventing overheating, water boiling and reducing scaling problems. Also all products of combustion are cleared from the combustion chamber which is filled with fresh air. The fan usually continues for 2–10 minutes and is often controlled by the printed circuit board electronics or by a thermistor in the flue or on the flow pipe away from the heat exchanger until water temperatures are down to 60°C.

Either method prevents the latent heat contained within the heat exchange from boiling its contents. If this happened the boiling water would cause the overheat thermostat to operate, shutting down the combi boiler to lockout. The combi boiler would not begin its ignition sequence until it has been manually reset.

Fault Finding Over Run
Occasionally the pump or fan over run thermostat fails. It is usually calibrated to be 'on' when the water is 60°C and 'off' below this. Test it is working by checking the continuity when the water is hot. It should read less than 1ohm. When the water is cool (less than 60°C) it should read infinity or overload (OL).

To establish there is an over run fault (fan or pump):-

1. Reset combi boiler if necessary
2. With the central heating turned off create a demand for heat by turning on a hot tap
3. Allow tap to run for 2 or 3 minutes
4. Stop heat demand by closing the tap
5. Wait 30 seconds and create demand by opening hot tap
6. If the combi boiler does not fire has gone to lockout and needs to be reset, there is a fault with either the fan/pump over run or the over run thermostat *(see Thermostats – 6.21)*

Remember, a heat exchanger which is heavily sludged or badly scaled may only overheat its contents when there is no/slow water flow. This would again activate the overheat thermostat.

Also, a fan or pump that is beginning to fail, but still running, may be ineffective in over run mode causing the combi boiler to lockout.

Fault Finding Fan
The main components of a flue fan are the motor, bearings, impellor, electrical connections and air pressure hose ports and all can fail. A failed/failing fan will:-

(a) Not operate the air pressure switch or activate thermistors
(b) Turn off the combi boiler and it may go to lockout
(c) Cause permanent pilot flame to extinguish
(d) Not allow main burner to light
(e) Possibly be noisy

On occasions there can be 'strange' fan/air pressure switch faults and may be due to the air pressure hose connection ports partially or wholly blocked.

Air pressure hose connection ports partially blocked.

1. Dismantle flue canopy
2. Remove the Venturi of the fan
3. Inspect air ports. There is often a whitish powder deposited and usually in the negative port
4. Clean air ports. The port may be brass, if plastic be careful as it may break whilst cleaning it

Fan Stopped/Will not Start

Reason	Repair
Domestic hot water switch	Set demand and check continuity. Less than 1 ohm O.K. or replace – *6.20*
Pump failing/failed	Check pump, repair/replace – *6.18*
Flow switch faulty	Check/replace flow switch – *6.10*
No power from printed circuit board	Check for 230V at fan terminals – *6.11*
Air pressure failed at normally open	Check air pressure switch repair or replace – *6.1*
Motor winding burnt out	Check resistance or coil. If faulty replace fan – *6.11*
Bearings wearing out	Check balance and feel/Replace fan – *6.11*
Rotor jammed	Free rotor if possible or replace fan – *6.11*

Fan Runs Slowly/Noisily

Reason	Repair
Fan unbalanced	Clean away dirt and fluff – *6.11*
Bearings wearing	Lubricate bearings/replace fan – *6.11*
Fan slipping on shaft	Tighten grub screw – *6.11*

Combi Boiler Fires O.K. when Demand Stops, Goes to Lockout

Reason	Repair
Fan or pump over-run has failed, overheat thermostat operated, combi goes to lockout	Check fan and/or pump O.K. also check overheat thermostat

Service and Repair

If the fan will not start, establish demand for heat from first domestic hot water and then central heating. Remember to set programmer to on and the thermostat to call for heat. Using a multimeter set on volts a.c., test for presence of 230V at fan terminals. If zero reading check printed circuit board and electric supply *(see Electronic Boards – 6.6)*. The fault may be overheat thermostat, boiler thermostat, low pressure and flow switches. Isolate the appliance and remove the connections at the fan terminals. Using multimeter set on Ohms (Ω) measure the coil resistance/continuity at the terminals:-

Flue fan coil

Less than 1 Ω	=	Fan coil burnt out or short circuit. Replace fan
Overload/infinity	=	Fan is faulty. Replace fan
Single speed fan	=	30–100 Ω O.K.
Double speed fan	=	Low speed 200+ Ω O.K. High speed 30–100 Ω O.K.

If the fan coil is okay it may be the rotor is jammed or the bearings are worn out. Remove fan and check for obstructions. Check free running and balance of fan. Remove dust and debris from impellor.

A noisy fan is usually an indication that the fan is failing. Always replace the fan before complete breakdown occurs. On some fans it is possible to lubricate or replace the bearings and/or replace the rotor.

6.11 Flue Fans

Faults with the fan (and flue way) may give rise to the following faults:-

(a) Pilot flame will not remain alight
(b) No ignition sparking sequence
(c) Main burner will not fire
(d) Noise
(e) Lockout during fan 'overun' period

Fan Resistor

Fan speed can be controlled in various ways including electronically via the printed circuit board, a step down transformer or a resistor. As resistors generate heat it follows that the larger the resister the more heat is generated. Printed circuit boards are heat sensitive so the fan power resistor is usually enclosed in a metal casing fitted to a heat sink attached to the combi boiler chassis. The heat sink dissipates heat from the resistor to the chassis preventing the resistor overheating.

If the heat is not transferred away from the resistor it will overheat soldered points causing an open/bad joint, possibly affecting fan speed, so check connections. This overheat of the power resistor may cause the resistor to burn out. To check the integrity of the resistor:-

1. Disconnect resistor
2. Set multimeter to Ohms (Ω)
3. Test resistance across ends:-

Overload/infinity = Resistor burnt out

Introduction

Gas valves control the supply of gas to the burner in response to control signals from the printed circuit board. They provide extremely accurate gas flow rates to match the requirements of the boiler load.

Gas valve – two solenoids

There are many variations of combi-boiler gas valves. From the traditional multi-functional valve used in conjunction with a pilot burner, thermocouple and room/boiler thermostat to combined valve and ignition systems for direct spark ignition (DSI) or intermittent pilot (IP).

Design and Construction

Usually made with a die cast aluminium body the multi-functional gas valve is made up of different components contained within the body or as a bolt-on. They will usually consist of a filter, thermoelectric valve, solenoids, pilot regulation sensor and pressure tests points. Modulating combi-boilers with a permanent pilot flame may have 2 solenoids for main gas rate and modulating.

When electronic ignition or flame rectification is used in a combi-boiler the gas valve may have a combination of 3 solenoids:-

Pilot Solenoid
Main Solenoid
Modulating Solenoid

These solenoids may be energised by alternating current (a.c.) or direct current (d.c.) and by differing voltages.

Typically, the pilot and main solenoids are 230Va.c. with the modulating solenoid energising with 24Vd.c. Always read data label and manufacturer's instructions *(see Solenoids – 6.19)*.

Connection to the gas supply pipe and the gas pipe to the burner is often with screw on flanges or sealed with rubber 'O' rings.

Operation

There are several types of gas control valves fitted to combi boilers and the most common is the modulating gas valve, commonly known as a modureg. This gas control valve enables the modulation of gas rates by using an electrically operated solenoid *(see Solenoids – 6.19)*.

Typical gas valve

Replacing the manually set spring loading set up, the solenoid is normally fitted to the top of a constant pressure governor. The solenoid armature will increase/decrease the gas rate in line with a voltage supplied across the coil. The voltage is controlled by the printed circuit board via the thermistor and potentiometer. The temperature of the thermistor indirectly controls the amount of voltage the potentiometer sends to the solenoid. This arrangement allows the modulation of the burner pressure to be matched to the loading placed on the combi-boiler.

Whether the gas control valve is as above or variable air loading or a manufacturers bespoke design the basic principle is the same i.e. accurate and safe delivery of gas to the pilot/main burner.

6.12 Gas Valves

Fault Finding

The gas control valve must have a sufficient gas supply to operate the combi boiler at maximum demand and especially when in domestic hot water mode. When the central heating first calls for heat this requires the supply pipe work from the meter to be checked for correct sizing:-

1. Attach a manometer to the inlet pressure test point
2. Create a maximum demand by fully opening a domestic hot water draw off point
3. The manometer must read between 20–22 mbars for natural gas or 35–37 mbars for LPG

If the inlet pressure is correct, check the adjustment of the modureg to confirm it modulates within the boiler range:-

(a) Turn off and electrically isolate combi boiler
(b) Remove the connections at the modulating solenoid
(c) Slacken the screw at the burner pressure test point and attach a manometer
(d) Turn on combi boiler
(e) Create a demand for domestic hot water. The burner should light at minimum pressure
(f) Check the manometer reading of the minimum burner pressure is as specified by the manufacturer
(g) If adjustment is necessary hold the plastic nut still with a spanner and turn the brass nut under the plastic cover until the minimum pressure is as specified by the manufacturer:-

Clockwise to increase pressure

Anti-clockwise to decrease pressure

(h) Stop domestic hot water demand, turn off and electrically isolate the combi boiler
(i) Replace connectors to the modulating solenoid, ensuring they are clean and well fitting
(j) Turn on combi boiler
(k) Create a demand for domestic hot water. The burner should light at maximum pressure
(l) Check the manometer reading of the maximum burner pressure is as specified by the manufacturer
(m) If adjustment is necessary, hold the brass nut still with a spanner and turn the plastic nut until the maximum pressure is as specified by the manufacturer
(n) The adjustment of one nut may slightly affect the position of the other, so always re-check the minimum and maximum settings
(o) Remove manometer, tighten test point screw and test for gas soundness with LDF

Remember, if there is no voltage from the printed circuit board to the solenoid then the fault is before the gas control valve. Is the fan operating? If it is, the fault is before the fan. If there is correct voltage and the valve does not work, fit a replacement.

If the gas control valve is faulty and cannot be repaired, a like for like replacement must be fitted:-

(1) Turn off gas supply and combi boiler
(2) Electrically isolate combi boiler
(3) Remove all solenoid connections making a note of their position
(4) Undo all unions to and from the valve
(5) Remove any screws and supporting brackets
(6) Fit new valve in reverse order using new gaskets as appropriate
(7) Attach manometer to test point, i.e. new valve
(8) Switch on gas supply
(9) Check for gas soundness at all unions made
(10) Switch on electric and combi boiler
(11) Check the gas pressures are correct
(12) Remove manometer, tighten test point screw and check for gas soundness with LDF

Introduction

As with traditional/system central heating boilers all combi boilers have a main or primary heat exchanger to heat the central heating system water. Originally cast iron, heat exchangers are now of copper with condensing combi boiler heat exchangers made of stainless steel or aluminium. The waterways within the main heat exchanger contain central heating system water. This water is heated by the hot flue gases and not by direct contact of the flame from the main burner. If the flame did play directly onto the heat exchanger then this would constitute flame impingement which would prevent correct combustion and produce the highly toxic and dangerous gas carbon monoxide. Also, localised heating of the exchanger would cause the exchanger to 'fail' prematurely.

Primary heat exchanger

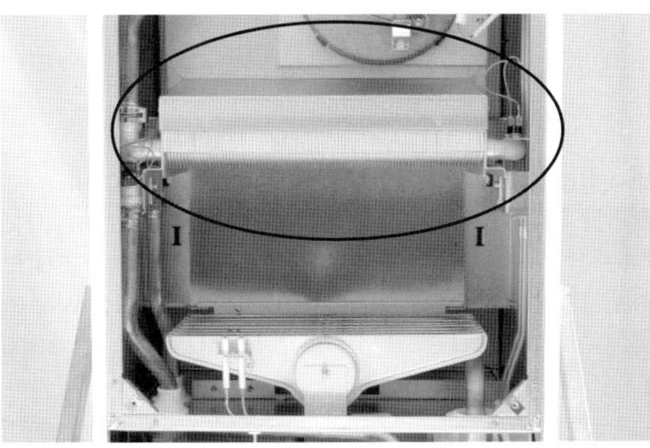

Primary heat exchanger positioned over burner

The hot flue gases heat the outside of the water way tubes by convection. Conduction moves this heat from the outside to the inside of the tube. This in turn heats the system water contained within the tubes by conduction. To make the exchange of heat more efficient, fins are attached to the tubing which increases surface area. These fins collect more heat from the hot flue gases and transfer the heat by conduction, to the water filled tubes.

Similar to modern system boilers the copper, aluminium or stainless steel primary heat exchangers have a low water content i.e. lower than the older and less efficient cast iron. The low water content principle 2–4 litres allows for rapid recovery rates. The consequence of this low water content is that when the boiler switches off residual heat within the primary heat exchanger must be dissipated to prevent overheating (kettling).

Low water content heat exchangers require a minimum flow rate of water passing through the waterway tubes to prevent boiling the water and/or damaging the heat exchanger. To facilitate this a fan and/or pump overrun is used and an automatic By-pass must be fitted *(see Automatic By-pass Valves – 6.3)*. Many combis are now manufactured with a built in pre-installed By-pass allowing the circulating pump to continue to run after the boiler switches off. The water contained within the heat exchanger will now be circulated and overheating will be prevented. This is known as pump over-run *(see Pumps – 6.18)*. Either way, always follow the manufacturer's installation instructions and fit an automatic by pass if required.

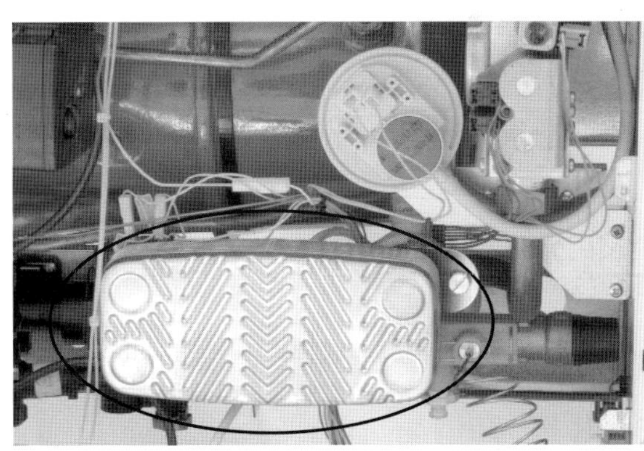

Domestic hot water heat exchanger

Domestic Hot Water Heat Exchanger

There are two ways of heating the domestic hot water within a combi boiler system; direct or indirect. In fact, combination boilers are frequently classified and sub-classified as direct or indirect *(see What is a Combination Boiler? – 1.1)*.

Direct Combi Boiler (Direct Hot Water Heat Exchanger)
Direct combi boilers have the domestic hot water pipes (referred to as secondary) running either inside or next to the central heating pipes within the primary heat exchanger. Some combi boilers use a directly heated secondary heat exchanger through which the domestic hot water flows. However, do not confuse this configuration with some models of condensing combi boilers.

Indirect Combi Boiler (Water to Water Heat Exchanger)
Indirect combi boilers are so called because the domestic hot water is not heated by the flue gases. Remember the main burner flame NEVER plays directly onto the heat exchanger – heat is always transferred to the water in the exchanger by the hot gases passing around it. The domestic hot water is heated by absorbing heat from the primary water heated and circulated in the main heat exchanger to the domestic hot water heat exchanger, also called a calorifier. Hence, the term 'indirect' or 'water to water heat exchanger.' There are two types of indirect water to water heat exchangers or calorifiers; the older and now less popular cylinder type and the now more popular plate type.

Cylindrical Type
This is now rarely used in combi boilers. However, it is similar to an indirect domestic hot water cylinder (in reverse) with a small cylinder tightly packed with a copper tube coil.

When a hot water draw off point is opened the mains supplied cold water flows through the coil. As it flows (usually entering the top and exiting from the bottom of the cylinder) it absorbs heat from the heated primary water within the cylinder and is delivered to the draw off point by main pressure. Unfortunately, to allow an efficient and usable transfer of heat the coil has a high resistance to water flow. This results in low flow rates, typically less than 10 litres per minute at a 35°C ΔT. Further, in hard water areas, scale formation can further reduce flow rates *(see Corrosion & Lime Scale – 2.4)*.

Plate To Plate Type
The plate type of domestic hot water heat exchanger is the most popular and nearly all new combi boilers now use this type. It is smaller than the cylindrical type and so the combi boiler overall size can be reduced making it more kitchen friendly. It is comprised of thin copper water-ways and stainless steel plates, giving a large surface area, promoting heat transfer and increasing efficiency with the primary hot water from the main heat exchanger flowing in an opposite direction to the mains cold water flow. This design increases heat uptake (and therefore efficiency) and further design features produces water turbulence which further increases efficiency and reduces scale build up. The smooth copper water ways expand and contract with heating and cooling. This movement helps to prevent the build up of scale which tends to break away from the tube wall.

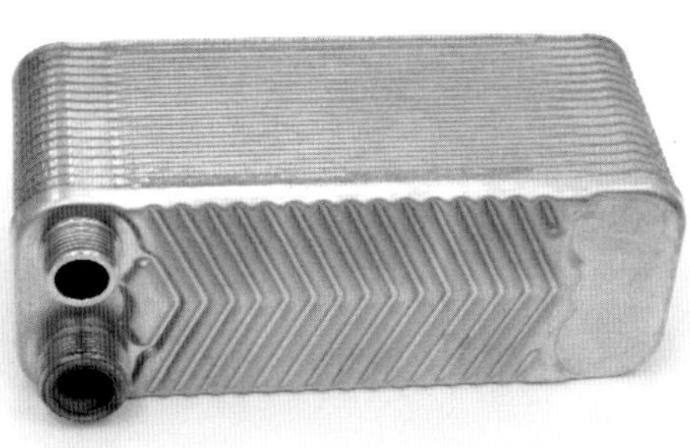

Plate to plate domestic hot water heat exchanger

The plate type heat exchanger can deliver high flow rates, typically up to 16 litres per minute at 35°C ΔT.

Fault Finding

Main Heat Exchanger
The main heat exchanger can only have 2 apparent faults:-

1. **Leak** – at unions or in body of heat exchanger
2. **Noise** – 'kettling' due to sludge/scale, a fan or pump overrun fault, By-pass faulty, limit thermostat faulty, etc

Leak
There may be a system leak which will be evident by a frequent drop in system pressure *(see Pressure Gauges – 6.17)*. This fault is diagnosed by having to regularly top up the system using the filling loop *(see Filling Loops – 6.8)*. Serious central heating system leaks are obvious but smaller leaks are more

difficult to find. If these smaller leaks are in less obvious places (under floorboards etc) the primary system water may be evaporating off the hot flow or return pipe. Difficult to find, the leak must be found and corrected or the problem will continue. Once the leak is repaired drain, flush and protect the system. If the leak is at the heat exchanger unions, tighten. If the leak is in the body of the heat exchanger it cannot be repaired and must be replaced or the combi boiler replaced.

Sludge/Scale Build-up
As combination boilers are sealed systems, sludge build up is generally not a problem unless there is a system fault. Scaling within the main heat exchanger reduces the primary water flow rate through it as the water ways cross sectional area are reduced. The consequences of this is the reduced circulating primary water does not take enough heat away from the exchanger and may activate the boiler thermostat causing the burner to shut down and then restart (cycling). Radiators and hot water may not reach temperature and the domestic hot water may flow hot and cold. Alternatively, the over-heat thermostat may be activated, shutting down the combi boiler, possibly to lockout.

As the water starts to boil within the main heat exchanger, due to hot spots, the boiler may rattle or make a 'kettle' noise. A further consequence of scaling is it prevents the efficient heat transference from the fins to the waterways which not only prevents the primary water from efficiently heating but causes the fins and outer walls of the heat exchanger to overheat. This overheating can distort or damage the fins. Hopefully, the boiler thermostat or overheat thermostat will operate and shut down the combi boiler before damage is caused.

In hard water areas fitting a scale reducer can help against scale build up, however, ensure the manufacturer recommends and supports this method of scale reduction. A central heating system which is drawing in air may be corroding, there may be electrolytic corrosion, dezincification type corrosion or heavily chlorinated water may be contributing to system corrosion. Any of these factors may cause a sludge build up in the main heat exchanger causing similar symptoms as scale build up. Drain, flush and protect the system with corrosion inhibitor *(see Flushing & Protecting – 2.5)*.

Fault	**Reason**	**Repair**
Reduced heat output to domestic hot water and central heating	Sludge/scale build up	Drain, flush and protect – *2.5*
Boiler cycling	Sludge/scale build up	Drain, flush and protect – *2.5*
	Boiler thermostat failed	Replace boiler thermostat – *6.21*
Reduced domestic hot Water temperature	Sludge/scale build up	Drain, flush and protect – *2.5*
Fins melted/distorted	Sludge/scale build up	Replace and/or drain, flush and protect – *2.5*
	Main flame playing on heat exchanger	Check position of main burner and check gas rates for over gassing – *5.8*
Noise (kettling)	Sludge/scale build up	Drain flush and protect – *2.5*
		Replace boiler thermostat – *6.21*
		Service/replace pump – *6.18*
Leak	Corrosion or crack caused by overheating	Replace heat exchanger – *6.13*

Fault Finding

Domestic Hot Water Heat Exchanger (Calorifier)

This can have the same 2 apparent faults as the main heat exchanger.

Leaks

If the domestic hot water heat exchanger is leaking it must be replaced as it is not repairable. Should the leak be across a common wall within the heat exchanger the system pressure will rise and show at the pressure gauge and may cause the pressure relief valve to discharge.

Sludge/Scale Build Up

As the cold supply water is heated, scale builds up in the waterways. Initially this may show as the domestic hot water not hot enough. This further restricts the domestic hot water flow at the draw off point. The reduced heat transference efficiency from the primary water causes 'cycling' on the limit thermostat and alternating hot/cold water. It may overheat and cause the overheat thermostat to shut down the boiler (lock out).

Modern calorifier waterway walls are now coated to help prevent scale build up. Sludge build up faults can also develop in the domestic hot water heat exchanger. The flow channels are narrow and sludge can easily be trapped. The effects of sludge is similar to scale formation.

Fault	Reason	Repair
Hot water not hot enough	Sludge/scale build up	Drain, flush and protect system or replace calorifier
Hot water alternating hot and cold	Sludge/scale build up	Drain, flush and protect system or replace calorifier
System pressure rise, possible pressure relief valve discharge	Mains cold water crossing common wall to primary circuit	Replace calorifier
System pressure rise, possible pressure relief valve discharge	Filling loop letting by	Close isolators and remove filling loop – *6.8*

Service and Repair

The main heat exchanger is serviced in slightly different ways depending on the material used in its construction. The objective in all cases is to remove any loose debris and to ensure the heat exchanger fins are clear of any obstruction. Use a soft brush and vacuum cleaner but be careful not to distort any of the fins (blades). Never use chemical products or steel brushes on combi boiler heat exchangers.

Some modern condensing combi boilers using alloy/stainless steel heat exchangers require a cleaning blade and bespoke brush. Condensing boilers require the condensate collector to cleaned.

Unions should be tightened if leaking but always follow the manufacturer's instructions. If the heat exchanger is leaking it is not repairable and must be replaced. Should the domestic hot water heat exchanger be leaking across the cold water supply and primary circuit face the system pressure will rise and may discharge through the pressure relief valve *(see Pressure Relief Valves – 6.16)*.

The domestic hot water heat exchanger should have any debris carefully removed and be checked for leaks. Some combi boilers have a plastic plate to plate heat exchanger which when faulty, usually leaking, can be replaced with a more reliable copper type including the connecting tubes.

Introduction

There are essentially 3 ways to ignite the gas/air mixture within a domestic gas appliance:-

Naked Flame
Filament Ignition
Spark Ignition

The use of a hand held naked flame to light the pilot or main burner is a thing of the past.

Filament ignition systems have not been used within domestic gas appliances since the early 1970's, having been replaced by spark ignition systems. The filament is a small coil of thin resistance wire made of platinum, heated by passing a voltage of approximately 3V from a battery or mains electricity via a step down transformer. The filament head is positioned next to the pilot jet and arranged in such a way to allow air to the pilot flame.

Glowcoil ignition systems are similar to filament ignition systems but usually use mains electricity, via a step down transformer, which provides a voltage of approximately 2.5V and this passes through a heavier filament. Again, this is now obsolete.

Spark ignition is now the preferred method and is found in virtually all combi boilers. There are various ways to produce a spark but whichever spark method is used to ignite the gas the voltage must be between 5,000V and 15,000V. This voltage can be obtained by either:-

Piezo Electric Igniters – Permanent Pilot Flame
Mains Transformers
Electronic Pulse Systems – Intermittent

Piezo Electric Ignitors

Design and Construction

Only used on some early combi boilers in the late 1970's and early 1980's. The word piezo is from the Greek word piezein, meaning to press, so roughly translated would mean pressure and that is exactly how the spark is created by the ignitor device. It contains lead zirconate titanate (PZT) crystals which during manufacture are exposed to powerful electric fields. This polarises the molecules within the crystals. The ignitor device contains 2 crystals both about 12mm long x 6mm diameter placed in parallel but separated by a metal pressure pad. This pad is connected to the spark electrode via an ignition lead. The crystals are earthed to the casing of the ignition capsule. The ignition capsule is often a plastic moulded, water resistant device that requires no maintenance and is not serviceable.

Piezo ignition capsule

Operation

There are two types of piezo electric spark ignition systems used in older combi boilers, cam operated and impact (hammer) operated. Both types rely on pressure (stress or deformation) being appled to the crystals and when operated produce an electromotive force (emf) of approximately 6,000V between the 2 ends of the crystals. This is transmitted to the spark electrode via the electrode wire 2.3mm/2.8mm diameter, rated to 200°C, which is usually a thin wall PTFE or silicon rubber insulated cable, with a silver plated copper conductor rated to 15 kV.

The electrode, also known as a spark tip, has a high quality steatite and alumina insulator of approximately 6.35mm diameter and is available in different lengths, typically 31.75mm, 36mm, 44.5mm, 50mm and 60mm. The electrode is fixed in position using a collar with thread and nut or screw/rivet fitting.

When the lever of either type of ignitor is depressed pressure is applied to the crystals, the voltage builds up until the resistance of the spark gap (3mm) at the electrode tip is overcome and a high voltage spark is produced. When the lever is released the pressure is removed and another high voltage spark is produced.

Fault Finding

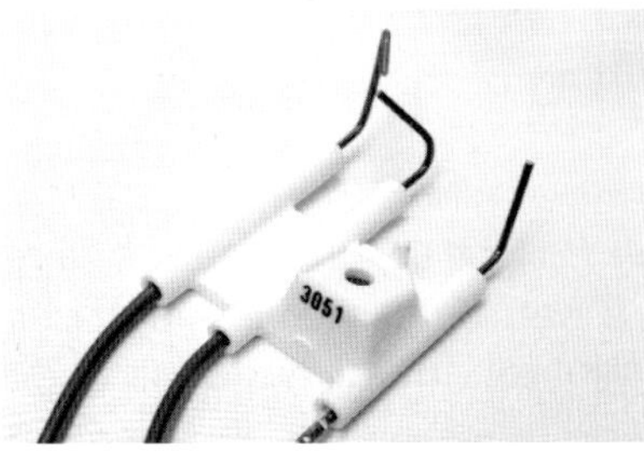

Spark tip and FSD combined

The crystals may become damp, possible due to condensation, causing the current to short out resulting in loss of spark at the electrode.

With all spark ignitors the correct position of the electrode (spark tip) is critical. The tip must be in a place where the air/gas mixture is within flammability limits and the gap must be of the correct size, usually 3mm – 4mm. Also, the speed of the air/gas mixture must be such that the flame produced will light onto the flame port. The earthing of crystals, the insulation of the electrode lead and the ceramic insulation of the spark electrode must be sound to prevent any leakage of the high voltage produced and required for appropriate sparking. Carefully check the terminal connections of the electrode lead and its insulation along its entire length.

Piezo Electric Ignitors - No Spark

Reason	Repair
Crystals damp	Replace ignition capsule/device – *6.14*
Crystals breaking down	Replace ignition capsule/device – *6.14*
Electrode lead terminal connection(s) unsound or disconnected	Clean/repair/remake connection – *6.14*
Electrode lead insulation damaged	Replace ignition electrode lead – *6.14*
Ignition electrode incorrectly positioned	Reposition ignition electrode – *6.14*
Ignition electrode tip pitted/damaged	Replace ignition electrode – *6.14*
Poor earth preventing complete circuit	Establish good earth – *6.14*

Service and Repair

As with all spark ignitors the electrode must be positioned correctly. Check the position gives an approximately 3mm – 4mm gap and the electrode tip is not badly corroded or damaged. To replace the electrode assembly electrically isolate the combi boiler, disconnect the cable and undo the 1 or 2 screws holding the bracket. Replace the electrode and re-assemble. If the ignition capsule/device is faulty disconnect the cable, undo retaining nut/clip and replace assembling in reverse order. Always follow the manufacturers instructions.

Mains Transformer

Used primarily with oil fired appliances the ignition step up transformer produces a voltage in a ratio of up to 1:100. Voltage of between 5,000V–10,000V are used with a 1:100 ratio producing 24,000V.

Electronic Pulse Ignition
Design and Construction

These kind of ignition systems use solid state electronics. A rectifier is used which converts an alternating current into a direct current (rectification). This rectified current passes to a capacitor. A capacitor consists of two thin parallel metal foils which are insulated from each other by waxed paper or a dielectric (insulator) such as mica. The now rectified direct current is connected across the metal foils and one becomes rich in electrons and the other proportionally poor. An external circuit connected to the capacitor will allow a current to flow (discharge) but only for an instant. Passing the current through a small step up transformer produces a very high voltage in the secondary winding which has the electrodes attached. This arrangement is capable of producing sparks at speeds of up to 8/second at voltages around 15,000V.

Operation

Electronic pulse ignition can be operated manually by a push button but usually the sparking discharge is controlled electrically by the printed circuit board.

Fault Finding

With all spark ignitors the correct position of the electrode (spark tip) is critical. The tip must be in a place where the air/gas mixture is within flammability limits and the gap must be of the correct size, usually 3mm – 4mm. Also, the speed of the air/gas mixture must be such that the flame produced will light onto the flame port. The insulation of the electrode lead and the ceramic insulation of the spark electrode must be sound to prevent any leakage of the high voltage produced and required for appropriate sparking. Carefully check the terminal connections of the electrode lead and its insulation along its entire length.

Electronic Pulse Ignition – No Spark

Reason	Repair
Electrode lead terminal connection(s) unsound or disconnected	Clean/repair/remake connection – *6.14*
Electrode lead insulation damaged	Replace ignition electrode lead – *6.14*
Ignition electrode incorrectly positioned	Reposition ignition electrode – *6.14*
Ignition electrode tip pitted/damaged	Replace ignition electrode – *6.14*
Polarity reversed	Check correct polarity at supply – *3.4*
Poor earth preventing complete circuit	Establish good earth – *3.4*

Re-Ignition Systems

Spark systems using electronics have the advantage of total electronic control. This means that ignitions systems can now be profiled differently. There are two such systems commonly used:-

Intermittent Pilot System

Operation

1. The system performs a safety check to confirm there is no flame present at the pilot assembly
2. If no flame is present the ignition sequence starts
3. The pilot solenoid is energised, allowing gas to pass to the pilot assembly
4. Sparking occurs
5. (i) If the gas is not ignited i.e. no flame is detected at the pilot, the pilot solenoid is de-energised and the combi boiler will go to lockout
   (ii) If the gas is ignited i.e. a flame is detected at the pilot, the main burner solenoid is energised allowing gas to pass to the main burner which will now fire, usually to full rate
6. If the flame is extinguished this will be detected, both solenoids will be de-energised and the above sequence will be repeated
7. If the flame is not re-established the pilot solenoid will be de-energised and the combi boiler will go to lockout

Note:- If the combi boiler goes to lockout it must be reset by pressing the reset button or by turning the boiler off and back on again.

Direct Spark Ignition System (DSI)

1. The system performs a safety check to confirm there is no flame present at the burner
2. If no flame is present the ignition sequence starts
3. The first of two solenoids will be energised, allowing gas to pass, usually at low rate, to the burner
4. Sparking occurs
5. (i) If the gas is not ignited i.e. no flame is detected at the burner, the solenoid is de-energised and the combi boiler will go to lockout OR it will recycle through the ignition sequence several times (as per the manufacturers settings) before eventually going to lockout
   (ii) If the gas is ignited i.e. a flame is detected at the burner, the second solenoid is energised allowing gas to pass at full rate to the burner
6. If the flame is extinguished this will be detected, both solenoids will be de-energised and the above sequence will be repeated
7. If the flame is not re-established the solenoid will be de-energised and the combi boiler will go to lockout

Fault Finding

Similar faults as described above can occur with both these ignition systems. Further, some combi boilers are sensitive to reversed polarity. Check for this at the incoming electrical supply *(see Electrical/Wiring – 3.4)*.

Service and Repair

Again, refer to above for service and repair. Note that most condensing combi boilers use this kind of automatic ignition and often the electrode assembly is supplied already mounted on a bracket. To replace, electrically isolate the combi boiler, remove the leads from the electrodes, unscrew the 2 (usually) fixing screws, ensure the spark gap on the replacement assembly is set correctly, typically 4.5mm (± 0.5mm), and re-assemble in reverse order.

Pilot Flames
Introduction

Older combi boilers may have a pilot flame as a means of igniting the main burner *(see Flame Supervision Devices – 6.9)*. It may be a permanent or intermittent type but either way gas is supplied to a pilot jet via a small injector. The size of this injector is critical to the correct burning of the pilot flame.

Remember, many combi boilers that are totally electronically controlled will not have a pilot assembly. The spark will go direct to the main burner and ignite the gas initially at low rate.

Fault Finding

There are several faults that may occur with the pilot flame with most of these faults as a direct result of other component failure.

If the pilot does not light it is usually as a result of component failure in the firing sequence BEFORE the pilot flame. If sparking occurs it can be assumed that the fan, air pressure hoses and switch (flue thermistor), gas valve solenoid(s) and ignition sequence and printed circuit board are operating O.K. Therefore, check gas supply and pressure, pilot injector for blockages and position.

If there is no sparking the fault can be anywhere in the firing sequence up to and including the ignition system.

It is logical to test the combi boiler in both central heating and domestic hot water mode. If it fires in one mode the fault will be much easier to trace *(see Fault Finding – 7)*.

Pilot Flame Too Big/Lifting Off

Reason	Repair
Excessive gas pressure	Adjust working gas pressure – *5.8*
Over sized pilot injector	Check and replace pilot injector – *6.14*
Incorrect adjustment	Check and adjust pilot injector – *6.14*
Fan failing	Check fan speed/operation and flue – *6.11*
Air pressure switch/air flow or hoses, thermistor faulty	Check air pressure switch and hoses or test flue thermistor – *6.1*

Pilot Flame Too Short

Reason	Repair
Insufficient gas pressure	Adjust working gas pressure – *5.8*
Partially blocked pilot injector	Clean pilot injector – *6.14*
Incorrect adjustment	Check and adjust pilot injector – *6.14*
Partially blocked pilot gas filter	Clean pilot gas filter – *6.14*
Partially blocked pilot supply tubes	Clean pilot supply tubes – *6.14*

Flame Extinguishing

Reason	Repair
Excessive draughts	Check flue and flue terminal are correct – *6.11*
Vitiated Air	Check flue and flue terminal for obstruction – *6.11*
Explosive ignition	Check armature in gas valve not stuck – *6.12*
Pilot flame incorrectly located	Correctly re-position pilot assembly – *6.14*
Pilot flame too short	Adjust working gas pressure – *5.8*

Service and Repair

Pilot assemblies should be checked and cleaned at service. Check the injector is securely assembled and seated squarely and jointing materials have not been used. If jointing materials have been used strip down assembly and thoroughly clean. Should the pilot not function correctly without jointing compounds replace the assembly.

Copper tube when heated and re-heated is prone to producing a flaking substance known as 'black dust.' Clear the pilot supply tubes by blowing through with compressed air using a can of air or a bicycle pump. Also clean the pilot injector by again blowing through. Avoid, if possible using a fine wire such as fuse wire to clear an obstruction in the pilot injector as the pilot orifice maybe accidentally enlarged.

6.15 Low Water Pressure Switches

Introduction

If the flow of primary water through the main heat exchanger is insufficient the heat exchanger may become overheated and damaged. The combi boiler will still go to shut down and the overheat thermostat will be activated at 90°C but damage may be caused to the exchanger.

To prevent possible damage a flow switch is incorporated to sense slow/no flow of pumped system water and a low pressure sensing switch that will shut down the combi boiler. Not all combi boilers are fitted with a low pressure switch.

The low pressure switch is sometimes known as a low system pressure switch.

Design and Construction

The low pressure switch consists of 2 parts. A pressure sensing device and a micro-switch with a 1/8" BSP screw thread fitting.

Operation

The low pressure switch receives 230V AC from the printed circuit board and if the pressure is OK (usually above 0.4 bar) it will return 230V AC to the printed circuit board. This input signal to the printed circuit board permits the printed circuit board to continue its sequence.

Fault Finding

Whichever switch is used within the combi boiler if the water pressure is low the combi will not fire in either central heating or domestic hot water mode. If the switch is faulty, again the combi will not fire in either mode. An air lock(s) moving around the central heating system, if not manually or automatically vented, will find its way to the low pressure switch and immediately the switch will sense low pressure and shut down the combi. Pump over-run or system water flow may move the air lock pass the switch so that when the householder/engineer attempts to start the boiler it fires! The air lock or further air trapped in the system eventually causes the same problem and the boiler shuts down again. The system must be vented of ALL air to prevent this problem re-occurring.

If the low pressure switch is not receiving 230V AC from the printed circuit board the fault is before the switch i.e. high limit thermostat, printed circuit board or electrical supply (including fuses).

The low pressure switch which senses the flow of primary water will detect:-

Low Water Pressure – A leak in the system or pressure lost over time will prevent the combi boiler from firing. The leak must be found and repaired and the system re-pressurised. If the low pressure is due to pressure lost over time 'top up' via the filling loop to about 1 bar.

Air Locks in System – An air lock will be sensed as low water pressure and a poorly vented system may cause many combi boiler shut downs.

Expansion Vessel Pressure Loss – If the diaphragm is split or the Schreader valve is leaking, system pressure will drop due to discharge at pressure relief valve

Combi Boiler Will Not Fire In Either Central Heating or Domestic Hot Water Mode

Reason	Repair
Low system pressure, lost through time	Re-pressurise System to 1 Bar – *6.8*
Faulty low pressure switch	Replace switch or use blanking kit (check pressure gauge is not faulty – *6.15*)
Pressure loss due to leak	Repair leak – pressurise system to 1 bar – *6.8*
Pressure loss due to overheat and water discharge through pressure relief valve	Establish cause of overheat – *6.16*
Pressure loss due to undersize/low charge/faulty expansion vessel	Recharge/replace vessel or fit additional expansion vessel – *6.10*
Air in system	Vent system at automatic/manual air vent(s), radiator vents and at pump vent screw – *6.2*

Service and Repair

Not all combi boilers are fitted with low pressure switches and rely on the high limit thermostat to protect the heat exchanger. These switches are not serviceable and if faulty should be replaced. The boiler needs to be drained down. They can be troublesome and some manufacturers now supply a blanking kit.

6.16 Pressure Relief Valves

Introduction

Pressure relief valve – twist type

When water is heated it expands. It expands by approximately 4% of its volume until 100°C is reached. As steam is formed above this temperature it expands rapidly. The central heating water contained within a sealed system will expand. Normal operating expansion is taken up by the expansion vessel mounted within the combi boiler or as a separate component installed close to the appliance *(see Expansion Vessels – 6.7).*

Should the central heating system water become over pressurised for any reason a safety valve set at 3 bar operates, discharging water from the central heating system preventing:-

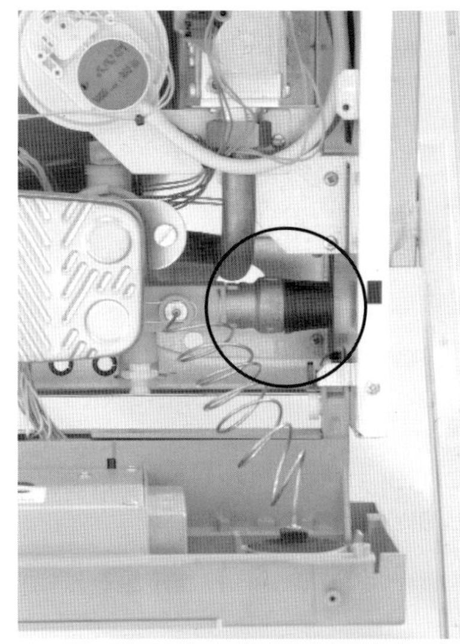

Pressure relief valve

(a) leaks at radiator joints
(b) leaks at system pipe work joints
(c) pipe work splitting
(d) radiator splitting
(e) boiler damage/failure
(f) system component damage/failure

A pressure relief valve, also known as a sealed system safety valve, is exactly that, a safety device.

Design and Construction

Pressure relief valves have a high discharge capacity and are always designed with a test mechanism, usually a twist type, with a red or black twist cap or a top mounted lever action. They must comply with standards and regulations and must be:-

(a) fitted close to the combi boiler
(b) fitted on the primary pipe work
(c) non-adjustable
(d) pre-set to operate at 3 bar pressure

Pressure relief valve – twist type

Most combi boilers have the pressure relief valve installed within the casing, usually factory set at 3 bar. Adjustable at manufacture to a pressure setting of 2 to 10 bar but tested to 15 bar. These robust components are generally available in two BSP sizes – $^{1}/_{2}$ inch and $^{3}/_{4}$ inch with a discharge capacity of 75 litres/minute and 150 litres/minute, respectively. Many combi boiler manufacturers specify and install more bespoke valves which may have a push fit drain connection, boiler drain, pressure gauge, capillary etc built into the body of the valve.

Pressure relief valve – lever type

Operation

The valve can be mounted vertically or horizontally either close to or within the combi boiler. The valve discharge outlet must point vertically downwards or horizontally. As the central heating system water is heated and expands causing an increase in pressure, the expansion vessel absorbs the increased water volume. If the system pressure should reach 3 bar the expansion relief valve will open allowing water to discharge. The valve is spring loaded and as the pressure reduces below 3 bar the valve will automatically close.

Discharge Pipe

Connected to the pressure relief valve outlet is the pipe to safely route discharged water. The connector/union is usually 1/2 inch (15mm discharge pipe) or 3/4 inch (22mm discharge pipe) BSP female but can be push-fit. In the event of a system failure and the pressure relief valve operating, hot water will be discharged. To prevent injury to people and animals or damage to property this discharge pipe must:-

(a) discharge in a visible place – so any discharge will be noticed
(b) discharge at low level where possible to prevent injury from discharging hot water
(c) at high level have discharge terminal turned into wall
(d) not discharge so as to damage property
(e) not be restricted or have any valve or component fitted in series – discharge route is safe and unobstructed to the atmosphere

Fault Finding

If water discharges from the pressure relief valve it is usually a symptom of a component fault in the combi boiler or possibly the central heating system. The pressure relief valve is a safety device and will discharge water if the system pressure exceeds 3 bar. The most common fault with the valve is leaking. Often this leaking starts after manual or automatic valve discharge:-

1. New installations, combi boiler replacement, work or extension on existing system and manual test at time of servicing can all give rise to the valve leaking due to rust, scale, debris etc sticking to the valve seating. This prevents the valve closing neatly on its seating with a resultant leak.
2. An old poorly maintained central heating system can result in the valve/valve seating perishing/corroding.

As it is rare for the pressure relief valve to 'fail' in its pressure control operation any discharge will be a symptom of another fault.

Valve Discharge at Fill-Up

Reason	Repair
System pressure greater than 3 bar due to overfill	Remove filling loop and reduce system pressure to 1 bar (cold) – *6.8*
Filling loop attached and letting by	Fully close isolators or replace and remove filling loop – *6.8*
Pressure gauge incorrectly reading less than 3 bar	Replace pressure gauge (like for like) – *6.17*
Pressure relief valve faulty (unlikely)	Replace pressure relief valve – *6.16*

Valve Discharge

Reason	Repair
Pump failing/failed, causing water in heat exchanger to boil	Check and repair/replace pump – *6.18*
Pump overrun failing/failed	Trace and repair fault at overrun thermostat/PCB – *6.18*
Expansion vessel under pressurised	Re-pressurise expansion vessel – *6.7*
Schreider valve leaking	Tighten or replace Schreider valve and re-pressurise or replace expansion vessel – *6.7*
Expansion vessel diaphragm split or porous	Check for water at schraeder valve when depressed. Replace expansion vessel – *6.7*

Expansion vessel under sized for installation	Replace with correct size vessel or fit additional expansion vessel – *6.7*
Gas rate too high	Check injector and measure gas rate/adjust burner pressures – *5.8*
Limit/overheat thermostat faulty	Replace thermostat – *6.21*
Obstruction preventing valve closing on seating	Clear obstruction or replace pressure relief valve – *6.16*
Domestic hot water heat exchanger leaking across to primary system	Replace domestic hot water heat exchanger – *6.13*
Attached filling loop letting by	Fully tighten isolators, remove filling loop and inform customer – *6.8*

Service and Repair

These valves should be checked for correct operation at the annual combi boiler service.

1. Is the pressure relief valve blow off pressure correct for combi boiler and is it non-adjustable?
2. Does the valve operate when manually tested. Turning the control cap or lifting the lever will lift the valve off its seating allowing water to discharge?
3. When the control cap is released ensure it re-seats correctly and does not drip *(see Fault Finding in this section)*
4. Does the discharge pipe fall continuously, have a clear and unobstructed route to the atmosphere and discharge safely and visibly?

The pressure relief has no serviceable parts and should it fail to operate correctly or is letting-by it must be replaced.

Always fit new pressure relief valves and ensure that the new valve is the same size and pressure rating as the original. The size and pressure rating of the new valve will be pre-set by the manufacturer and ready to fit to the combi boiler.

1. Electrically isolate combi boiler *(see Electrical Safety – 4.2)*
2. Drain down the combi boiler or central heating system or use a pipe freezer *(see Pipe Freezers 5.5)*
3. Pressure relief valves are held in place and attached with various retaining clips and connections e.g. pressure gauge and discharge pipe connections. Undo all associated connections
4. Remove the valve
5. Fit new valve
6. Reconnect in reverse order
7. Fill, vent and re-pressurise central heating system. If system was drained down add chemical inhibitor *(see Flushing & Protecting – 2.5)*

Should it prove to be impossible to remove the failed pressure relief valve it is possible to fit a replacement of the same size and pressure rating as the original. Fit in the ‘flow’ pipe work as close to the combi boiler as possible. New discharge pipe work must be installed and connected to original discharge or to a point that safely discharges. All discharge pipe work is to have a continuous ‘fall.’ A pressure gauge must also be installed *(see Pressure Gauges – 6.17)*.

Pressure relief valve – twist type variation

Introduction

The pressure the atmosphere exerts at sea level is referred to as atmospheric pressure (15 lb/sq"). This pressure is referred to as 1 bar. The same increment again is 2 bar and so on.

Sealed system combi boilers are designed to have a normal working pressure of between 1.0 – 2.5 bar. Combi boilers will generally operate within a system pressure range of between 0.4 – 2.9 bar. At 3 bar the pressure relief valve (factory set at 3 bar) will operate and discharge system water *(see Pressure Relief Valves – 6.16)*. As the pressure of a sealed system is critical to its correct functioning there must be a readout (gauge) available of that pressure.

Pressure gauges should be in a visible and accessible position and so readable from the system filling position. This allows installer, customers and service engineers easy access to check system pressure and to monitor the pressure when charging the system using the filling loop.

The pressure gauge is a much under-used diagnostic instrument which can indicate many faults to the informed engineer.

Pressure gauge mounted in facia

Design and Construction

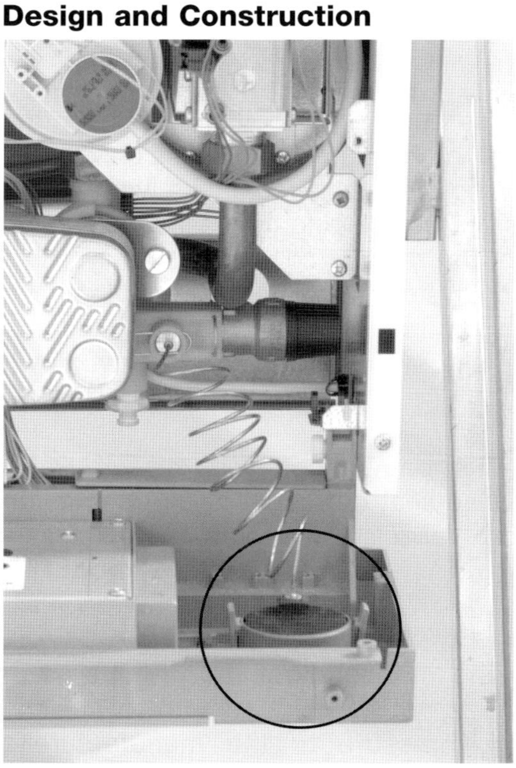

Rear of pressure gauge showing capillary tube

The pressure gauge is normally pre-installed in combination boilers and generally has a range of 0–4 bar. The system pressure is transmitted to the bellows of the gauge by a capillary tube from pressure sensor located on the combi boiler pipe work.

Many pressure gauges have a pre-set indicator for the 'COLD' fill pressure and 'Blow Off' pressure indicator, normally set at 3 bar.

Operation

The system pressure is increased by topping up the system from the cold mains via the filling loop *(see Filling Loops – 6.8)*. Different combi boilers have a specified COLD pressure and manufacturers instructions should be followed. In the absence of instructions the pressure should be increased until the gauge reads approximately 1 BAR WHEN COLD.

As the system is heated the water expands and the system pressure increases. This increase is transmitted from the pressure sensing head via the pressure gauge capillary to the pressure gauge dial.

In a correctly designed system (correctly sized expansion vessel, number and size of radiators etc) the heated working pressure is in the range of 1 – 2 bar, again refer to manufacturers instructions.

6.17 Pressure Gauges

Fault Finding

If a combi boiler has failed or is not working and the pressure gauge reads between zero and 0.4 bar DO NOT ATTEMPT TO RESTART THE COMBI BOILER. Should the boiler fire the heat exchanger could be damaged due to lack of water. Check the following:-

1. **Is the pressure gauge working correctly?** Using the filling loop, charge the system for 3–4 seconds and observe the gauge for movement. If there is no movement (and no leak!) the pressure gauge is faulty and must be replaced. Be aware that if the gauge is faulty, by charging the system, the pressure will be increased and may cause a discharge through the pressure relief valve *(see Pressure Relief Valves – 6.16)*. This discharge can happen at the time of filling or subsequently when the system is heated. Reduce system pressure to 1 bar or manufacturers specification watching the gauge for the appropriate movement.

2. **Are there system leaks?** Inspect system including pump unions, radiators and radiator valve unions, pipe work, zone valve unions, drain-cocks, boiler connections, heat exchanger, discharged terminal, filling loop connector isolator etc.

Other than 1 above, a very high or very low gauge reading will be the indicator that there is a fault or fault developing within the combi boiler or if components are installed separately, the expansion vessel, pump, automatic air vent or By-pass valve.

Pressure Gauge Zero/Low

System Slowly Lost Charge – Water should of course not escape from any central heating system. However, micro leaks can cause a gradual pressure loss, until after several months, the pressure is low. This is either casually observed at the pressure gauge or is highlighted by the combi boiler not firing or shutting down *(see Filling Loops – 6.8)*.

Excessive Venting Of Radiators – Faults within the central heating system may be causing air to be sucked in to the system or the production of gases due to corrosion or electrolysis. These gases may travel to the top of radiators causing the typical 'cold/luke warm at the top' symptom. The radiator(s) is vented allowing the gas(es) to escape and the water to fill to the top of the radiator. This temporarily cures the 'cold top' radiator, but at each venting water escapes reducing system pressure necessitating fresh water top-up via the filling loop. The cause of air ingress or gas production must be established and corrected quickly before serious corrosion/electrolysis causes major system/combi boiler breakdown *(see 'Central Heating – Fault Finding & Repair' book)*.

Pressure Relief Valve Faulty/Leaking – If the valve is faulty it must be replaced *(see Pressure Relief Valves – 6.16)*. Should the valve be leaking it will be due to valve seat corrosion or an obstruction preventing the valve closing on the seat. In both cases fit a new valve *(see Pressure Relief Valves – 6.16)*.

Pressure Gauge Reads High

Excessive Fill Pressure – The householder may have over filled/pressurised the system and is not aware of how to reduce the pressure. Simply, reduce the pressure to the manufacturer's recommendation or in the absence of manufacturers instructions to 1 bar by operating the pressure relief valve or draining at a drain cock (found at the lowest point of the central heating pipe work).

Filling Loop Attached – The filling loop should be removed when not in immediate use. If left attached and the isolators are not fully closed or are letting by, water at mains pressure will enter the system causing a higher than required system pressure. This may also be accompanied by the occasional discharge from the pressure relief valve especially on heating. Fully close isolators, disconnect the filling loop and reduce the pressure to about 1 bar when cold *(see Filling Loops – 6.8)*.

Domestic Hot Water Heat Exchanger Leaking – A leak in a common wall of the domestic hot water heat exchanger will allow the cold water, at mains pressure, to cross over to the system water. The system water pressure will increase and show on the pressure gauge. As this fault gets worse the pressure relief valve may discharge. Check/replace the domestic hot water heat exchanger *(see Heat Exchangers – 6.13)*.

Pressure Gauge Reads Low/Zero When Cold and High/Discharge When Hot

Undersized Expansion Vessel – Noticeably after new install or extension to a central heating system. The increased volume/pressure created by the heated system is accommodated by the expansion vessel. If the vessel is undersized the pressure may increase to 3 bar causing the pressure relief valve to operate and discharge water.

The combi boiler will probably continue to operate in both domestic hot water and central heating modes. However, when the demand for heat ceases the system will cool, the pressure reduces to nearly zero and air may be drawn into the system. Topping up the cold system to recommended pressure (about 1 bar) and creating a demand for heat will start the cycle again i.e.

System cold – OK
System warms – pressure gauge reading increases
System gets hotter – pressure gauge reads 3 bar
Pressure Relief Valve discharges..................

This fault can cause other combi boiler faults. Do not operate the system until an additional expansion vessel, 5 or 8 litre should suffice, is fitted *(see Expansion Vessels – 6.7)*.

Expansion Vessel Pre-Charge Pressure Low – When installing a new combi or replacing/fitting a new expansion vessel always check the vessel pre-charge pressure is at the manufacturers recommended pressure (about 1 bar). The pre-charge pressure must be measured when there is no external pressure acting on it. The combi boiler must not contain water so test vessel before installing and also test any replacement/additional vessel before fitting *(see Expansion Vessels – 6.7)*.

Expansion Vessel Pressure Low Due To Faulty Schraeder Valve – Remove dust cap and tighten schraeder valve. Pressurise vessel and using leak detector fluid check around the valve for leaks. If the schraeder valve is leaking, replace valve, re-pressurise and test again. Some valves are not removable so the expansion vessel will need to be replaced *(see Expansion Vessels – 6.7)*.

Expansion Vessel Pressure Low Due To Faulty Neoprene Diaphragm – A split or old and porous diaphragm is usually diagnosed by water escaping through the schraeder valve when the valve is pressed. The heating system pressure increases until the pressure relief valve discharge as above. The diaphragm cannot be accessed and therefore a new expansion vessel must be fitted *(see Expansion Vessels – 6.7)*.

Manual or Automatic By Pass Faulty (or not fitted!) – As thermostatic radiator valves close down on radiators the flow resistance increases and the system pressure may also increase. A wrongly adjusted manual or faulty automatic By-pass valve can cause an overheat situation with possible pressure relief valve discharge and combi boiler shut down often associated with lockout *(see Automatic By-pass Valves – 6.3)* By-pass valves must:-

(a) be installed correctly
(b) be set correctly
(c) operate correctly

Pump Is Failing/Failed – If the pump slows or stops the system water flow slows down (or stops) causing the water contained in the heat exchanger to overheat. This may lead to kettling, cycling, boiler shut down, lockout and often an increase in system pressure indicated by the pressure gauge *(see Pumps – 6.18)*. If the discharge occurs when boiler firing ceases, the fault may be with pump or fan overrun.

Pressure Gauge Reads Zero/Low

Reason	Repair
System Leak	Inspect system for leak(s) – *6.17*
Faulty Gauge	Check/replace gauge – *6.17*
System slowly lost charge through natural losses	Re-pressurise system – *6.8*
Excessive venting of radiators by householder	Establish reason for air in system, trace reason, correct and re-pressurise system – *6.8 (see also 'Central Heating – Fault Finding & Repair' book)*
Pressure relief valve faulty/leaking due to obstruction at seating	Clear obstruction or replace pressure relief valve – *6.16*

Pressure Gauge Reads High

Reason	Repair
Excessive fill pressure	Reduce pressure to 1 bar (cold) by releasing water from system – *6.8*
Filling loop attached and isolators letting by	Fully tighten isolators, remove filling loop and inform customer – *6.8*
Domestic hot water heat exchanger leaking across to primary system	Replace domestic hot water heat exchanger – *6.13*
Faulty gauge	Check/replace pressure gauge – *6.17*

Pressure Gauge Reads Zero/Low When Cold and High/Discharge When Hot

Reason	Repair
Undersized expansion vessel after combi replacement/central heating extension	Replace with larger or fit additional expansion vessel – *6.7*
Expansion vessel pre-charge pressure low	Re-charge expansion vessel – *6.7*
Expansion vessel pressure low due to faulty schraeder valve	Tighten/replace valve and re-pressurise or replace expansion vessel – *6.7*
Expansion vessel pressure low due to faulty neoprene diaphragm	Replace expansion vessel – *6.7*

Manual or automatic by pass not fitted/set	Reset/replace by pass valve – *6.3*
Pump is failing/failed causing water to overheat in heat exchanger	Repair/replace pump – *6.18*
Pump or fan over-run faulty	Check pump – *6.18*. Check fan – *6.11*

Servicing and Repair

The pressure gauge, including the capillary tubes and sensing head, need little maintenance and if faulty should be replaced on a like for like basis. The gauge is held in place by screws, clips or lugs. The sensing head is often secured in position by a slide in clip. Handle the capillary tube carefully and ensure it is not crimped by fascia or other combi boiler fittings.

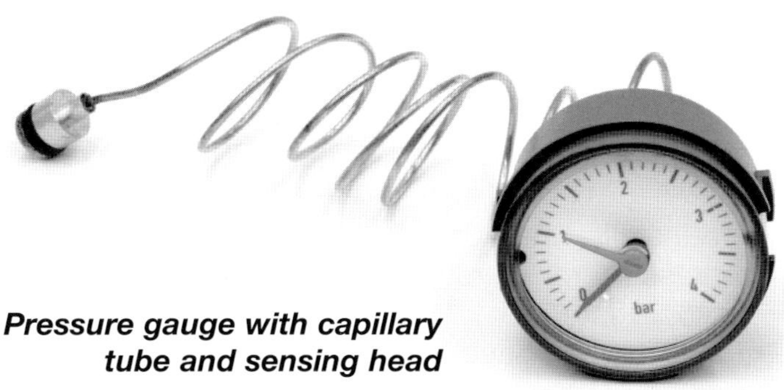

Pressure gauge with capillary tube and sensing head

Introduction

Circulating pump installed within combi boiler casing

When a combi boiler fires the water in the heat exchanger will be quickly heated. With low water content heat exchangers the system water is rapidly heated to temperature. This heated water is needed to meet a demand from the central heating or a domestic hot water requirement. The most efficient way of moving this heated water to either the central heating system or to the secondary heat exchanger is by using a machine. This machine will circulate the heated water to where it is needed so it can give up its heat to meet the demand. This machine is commonly referred to as a pump and less commonly as a circulator or an accelerator. To cover most situations the phrase 'domestic circulating pump' has been popularised.

Design & Construction

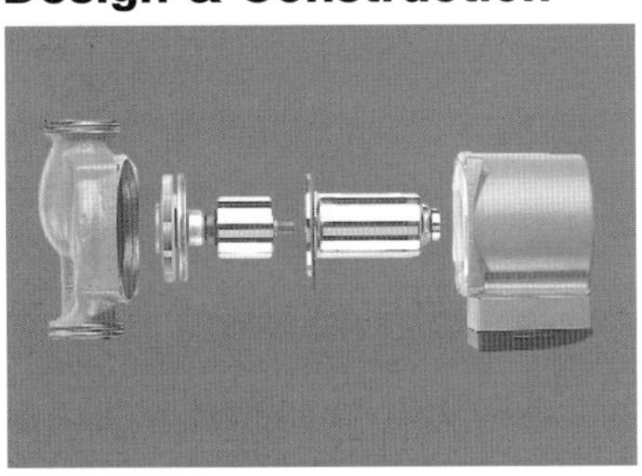

Circulating pump exploded view

Modern quality pump construction consists of a cast iron housing, stainless steel rotor can and rotor cladding with ceramic upper and lower bearings, a carbon based thrust bearing and rubber 'O' rings. The terminal box is constructed of plastics. The electric motor rotates the shaft at speeds of between 700 – 2000 rpm.

Circulating pump cross sectional view

Pre-installed pumps can operate within a pressure range of 0.1 – 10 bar (being factory tested to 15 bar) although most combi boilers have a pre-set maximum working pressure of 3 bar. The pumps are designed to operate at temperatures of typically 10 – 110°C so they can easily accommodate most conditions. In most combi boilers the pump is installed on the central heating return pipe.

The water contained within the pump both lubricates the bearing and keeps the pump motor temperature within its designed operating range. These pumps are robust and corrosion resistant.

3 speed circulating pump

Many older combi boilers are fitted with a fixed head circulating pump. This operates at one speed and has no adjustment. Its operating tolerances against pressure (head) is wide so suits many and varied systems. Unfortunately it is not especially efficient and can cause system noise.

Variable speed modulating circulating pump

In many modern condensing combi boilers an automatic variable speed circulating pump is fitted. This automatically adjusts its speed to meet the pressure fluctuations caused by differing heating demands. As thermostatic radiator valves close in response to set room temperatures the pump speed automatically adjusts, reducing the differential pressure as the resistance of the system rises. This saves energy (electricity) and reduces the possibility of system noise such as radiator valve noise and pipe flow noise. When they are fitted the manufacturer may also fit a by pass, therefore DO NOT fit an automatic by pass valve. Always follow the manufacturer's instructions.

Circulating pumps installed by the manufacturer within the combi boiler casing require a 230V mains supply. These pumps have a low power consumption ranging between 35 – 100 watts, (similar to a standard light bulb). Pre-installed pumps are electrically protected by the combi boiler fuse normally rated at 3 amps. The pumps are always earthed by the combi boiler manufacturer. DO NOT remove the terminal box cover without first disconnecting the electric power supply and carrying out electrical safety procedures *(see Electrical Safety – 4.2)*.

Operation

View of pump impellor showing vanes

As the motor rotates the shaft a circular veined wheel, known as an impellor, draws water in through its centre (eye) and throws it out at its edges by centrifugal force. As the water is thrown outward it creates a pressure drop causing more water to be drawn in at the eye. The faster the impellor spins the greater the centrifugal force created, which increases the water pressure in the system.

End view of pump impellor

Pump Overrun

This is a time or temperature device which enables the pump to continue running for a short time after the boiler has shut down. This facility allows the pump to continue circulating the water *(see Automatic By-pass Valves – 6.3)* through the pipe work removing the heated water from the combi boiler heat exchanger. If the water was not circulated the residual heat left in the exchanger would overheat the water, possibly causing it to boil. This could cause a noise (kettling) or cause the overheat thermostat to operate and send the boiler to lockout. Further, by preventing the water from over heating it reduces the potential for scale deposition.

By providing 230V to the pump when the water is over 60°C and after the combi boiler has shut down, the pump continues to run until the water drops below 60°C. The thermostat/sensor is normally closed when the water is above 60°C and opens when the water is below 60°C.

Fault Finding

On every occasion the combi boiler fires the circulating pump should run and when the boiler turns off the pump will either stop immediately or will run for a short time and then stop. When the pump runs on for a short while it will remove latent heat from the boiler. This is called pump 'over-run.' Circulating pump faults will either be:-

1. Pump failure
2. Pump speed low
3. Pump noise

Failure

There are four ways a circulating pump can fail to start:-

1. No electrical supply

1a. Faulty Capacitor

2. Worn out bearing(s)
3. Rotor seized
4. Burnt out motor windings

Electrical Supply (Faulty Capacitor)

If the pump does not start it may be there is no electrical supply going to the pump.

1. Remove terminal cover
2. CAUTION – using a multimeter check the voltage at the pump terminals. It should read 230V across the brown or red (live) and blue or black (neutral)
3. If there is no voltage the problem is at the combi boiler (check terminals for voltage), or if the fault is when the central heating is calling for heat the programmer or room thermostat is faulty. Check setting *(see 'Central Heating – Fault Finding & Repair' book)*
4. You can check the circulating pump has failed:-

   a) Turn off and electrically isolate the combi boiler
   b) Remove the three wires from the pump terminals making a note of positions
   c) Using a test lead fitted with a plug and 3Amp fuse, connect the three wires to the pump terminals (Brown-Live, Blue-Neutral, Green/Green and Yellow-Earth)
   d) Plug the test lead into a socket
   e) Switch the socket on and if the pump runs then the fault is not with the pump but with another component
   f) Remove the vent screw using the pump tool and using a 4mm screwdriver or the pump tool, quickly spin the rotor shaft. If the pump starts the capacitor is leaking/faulty. Replace pump or pump head

Note:- If the pump does not start immediately turn off the electrical supply or the pump motor (windings) may burn out.

Worn Out Bearings

If the bearings are worn out then the pump or pump head will need to be replaced. When checking for this be prepared for water spillage. Protect all electrics and electronics.

1. Turn off and electrically isolate the combi
2. Remove the vent plug in the face of the circulating pump using a pump tool or large blade screwdriver
3. If water pours out then the bearings are worn out. Mop up the spillage
4. Replace vent plug using pump tool or large blade screwdriver
5. Fit new pump or pump head

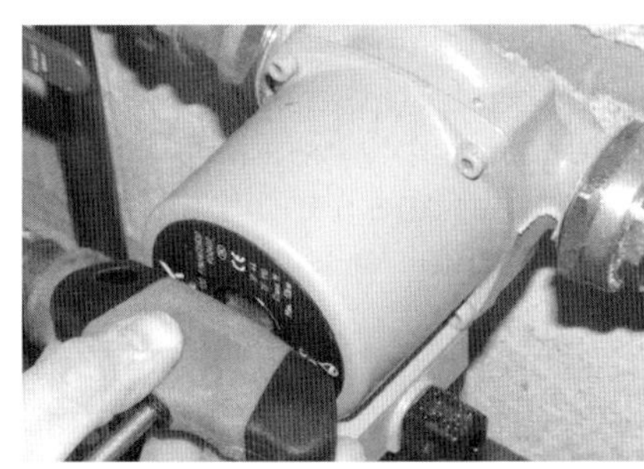

Pump venting using pump tool

Rotor Seized

Turning impellor using pump tool

The rotor may have seized due to particles of rust/sludge or scale preventing it from rotating. When checking for this be prepared for water spillage. Protect all electrics and electronics.

1. Turn off and electrically isolate the combi boiler
2. Remove the vent plug in the face of the circulating pump using pump tool. Mop up any water spillage
3. If just a few drops of water seep out fit a 3mm screwdriver (use pump tool) through the vent plug opening and gently turn the shaft anti-clockwise to dislodge particles
4. If the rotor shaft feels 'rough' as it spins or it will not move a new circulating pump or pump head is required
5. If the rotor shaft turns freely (anti-clockwise) replace the vent plug using pump tool, reconnect the power, turn the boiler on and check the pump is working correctly

Burnt Out Motor Windings

Having checked the electrical supply and there is 230V at the pump terminals, the rotor has been freed and now turns freely (anti-clockwise), but the pump does not work, then the motor has burnt out due to the rotor seizing and causing overheating in the motor windings. Often, when the motor windings do burn out the 3 amp fuse blows. Always check the fuse in the plug or fused connection unit AND any fuse within the combi boiler. To test and prove the motor windings have burnt out:-

1. Turn off and electrically isolate the combi boiler
2. Remove pump terminal cover and remove wires to pump terminals making a note of the connections
3. Use multimeter set on Ohms(Ω) and test the resistance across live and neutral terminals:-

Less than 1 Ohm = Motor windings burnt out – Replace pump or pump head

Mega Ohms/Infinity/Overload = Motor windings burnt out – Replace pump or pump head

120 – 200 Ohms = Motor windings O.K.

Pump Speed Low

To prevent kettling and overheating, the pump must run at speeds appropriate to the burner output level. If the pump does slow down due to sludge/rust/debris fouling the impellor or the pump is reaching the end of its useful life the following may be evident:-

1. Lack of heat output at radiators/domestic hot water
2. Boiler kettling
3. Frequent boiler shut down
4. Boiler cycling on limit thermostat

The debris fouling the impellor must be dislodged so the central heating system should be drained, flushed and protected *(see Flushing & Protecting – 2.5)*. If the pump is simply coming to the end of its working life then it must be replaced.

Pump Noise

The major cause of system noise is incorrect impellor speed. When the circulating pump speed is too slow the following may occur:-

Boiler 'kettling' due to water not passing over the heat exchanger fast enough causing it to become overheated.

Impellor

As with any central heating system component there is wear and tear. The circulating pump will give year after year of trouble free service but the bearings and rotor will gradually wear and lose their precision running ability. This 'out of true' state causes the impeller to run out of balance and create noise which is evident by the pump slightly moving when operating. This is a sign the pump is nearing the end of its working life and the only solution is to fit a new pump or pump head.

Air

When air enters the pump the noise can be very loud. It is relatively simple to vent the pump but the reason for the air entering the system must be established and corrected as dry running/cavitaiton of a circulating pump will damage it very quickly.

If the combi boiler is installed wrongly or at the wrong angle the top bearing may not receive system water to both lubricate and cool it. The pump will fail quickly if this is not corrected immediately. Use a spirit level to check the boiler is installed correctly.

Venting

Pump venting using pump tool

It is essential that when a new circulating pump has been installed, the combi boiler or the central system system has been drained the circulating pump is vented. It must be vented before it is allowed to start. This will remove air and allow system water to enter. The system water will cool the pump, preventing overheating and will also lubricate the shaft and bearings. Permanent damage may result from the pump being run dry even for a short time. ALWAYS vent.

Expect water spillage so have old towels/rags available to protect floors etc and to mop up.

1. Turn off and electrically isolate the combi boiler
2. Remove circulating pump terminal cover
3. Caution – using a multimeter check the voltage at the pump terminals. There should be a ZERO reading
4. Replace terminal cover
5. Remove the vent plug using a pump tool or wide blade screwdriver. Mop up water spillage
6. Fit a 3mm screwdriver (or use pump tool) through the vent plug opening and gently turn the shaft anti-clockwise
7. Rotate the shaft several times to remove any air
8. Replace vent plug. Do not over tighten
9. Replace any fuses and turn on the electrical supply
10. Set the room thermostat to maximum
11. Set the programmer to central heating 'on'
12. Turn on the combi boiler and check correct operation of pump and combi boiler
13. Reset room thermostat to original setting

Pump/Pump Head Replacement

When replacing a combi boiler pump there will be water spillage so be prepared to protect all electrics and electronics including flooring etc and have a bowl and old towels/rags to mop up.

1. Turn off and electrically isolate the combi boiler
2. Remove the circulating pump terminal cover
3. Caution – using a multimeter check the voltage at the pump terminals. There should be a ZERO reading
4. Note (write it down!) the colour of each wire and its corresponding terminal letter or number.
   Brown (or red) – Live to Terminal L or 1
   Blue (or black) – Neutral to Terminal N or 2
   Green/Yellow (or green) – earth to Terminal E or 3
5. Note (write it down!) the way the pump is installed
6. Note (write it down!) the water flow direction. This is indicated by an arrow on the pump often cast into the base of the pump
7. Remove each electric wire from its terminal
8. Drain down the central heating system or use a pipe freezer *(see Pipe Freezers – 5.5)*
9. Using a pump tool remove the vent plug and deal with any water spillage. Note the colour of the water. If it is clear with no debris the system water does not need to be changed. If it is dirty (a reddish or blackish colour) or contains debris the system MUST be flushed and protected *(see Flushing & Protecting – 2.5)*
10. Hold the circulating pump in one hand to steady it, loosen the union nut between the pipe work and pump with a pipe wrench, pliers or large spanner

Pump head fitting kit

10a. Alternatively, remove the pump head by using a pump tool (4mm Hex 'T' bar)
11. Holding the pump, completely undo both union nuts
12. Remove the circulating pump

Flushing and Protecting the System
It is essential to minimise the possibility of future damage or inefficient operation by ensuring the whole system is clean and protected against corrosion *(see Flushing and Protecting – 2.5)*

Fitting the New Pump
Circulating pumps should be replaced either on a like for like basis, if available, or leading manufacturers provide replacement guides and fitting kits/recommendations for using their pumps.

1. Do not apply sealants or jointing compounds to any surface including the threads of the isolating valves or the pump port threads. Assemble dry
2. With the flow arrow in the correct direction (check your notes) place the new circulating pump into position between the union nuts
3. Slide the new gaskets into place
4. Hand tighten both union nuts ensuring the gaskets do not move out of position
5. Steady the new circulating pump with one hand and tighten both union nuts. Do not over tighten. Ensure the pump is at the correct angles
6. Important – vent the circulating pump
7. Remove the terminal cover and using your notes rewire the new circulating pump
8. Replace the terminal cover
9. Replace any fuses and turn on the electrical supply
10. Set the room thermostat to maximum
11. Set the programmer to central heating 'on'
12. Turn on the combi boiler

Fault Finding
No Central Heating/No Hot Water

Reason	Repair
Pump failure – no electrical supply	Check electrical connections at pump – *6.18*
Pump failure – burnt out motor windings	Replace pump head – *6.18*
Pump failure – rotor seized due to rust/sludge sediment	Repair impellor movement or replace pump or pump head – *6.18*
Pump failure – worn out bearing(s) due to wear and tear	Replace pump head – *6.18*
Pump failed – dry running	Replace pump head and vent before turning on combi boiler – *6.18*
Pump failure – burnt out motor windings due to seized rotor	Replace pump/pump head – *6.18*
Pump capacitor faulty	Test capacity and replace if faulty or replace pump/pump head

No Central Heating/Hot Water O.K.

Reason	Repair
Room thermostat faulty so no signal to combi boiler	Check call for heat at room thermostat – see 'Central Heating – Fault Finding & Repair' book
Diverter valve faulty	Check diverter valve – *6.5*

No Hot Water/Central Heating O.K.

Reason	Repair
Diverter valve faulty	Check diverter valve – *6.5*

Insufficient Heat To Radiators

Reason	Repair
Pump speed slow due to sediment/sludge etc	Flush and protect system and overhaul pump – *2.5*

Frequent Combi Boiler Shut Down

Reason	Repair
Pump speed slow due to sediment causing combi limit thermostat to activate	Flush and protect system and overhaul pump – *2.5*

Pump Noise

Reason	Repair
Air in system/pump	Vent system and pump
Dry running of pump	Position combi boiler correctly
Impellor out of balance due to wear and tear	Replace pump head

Service & Repair

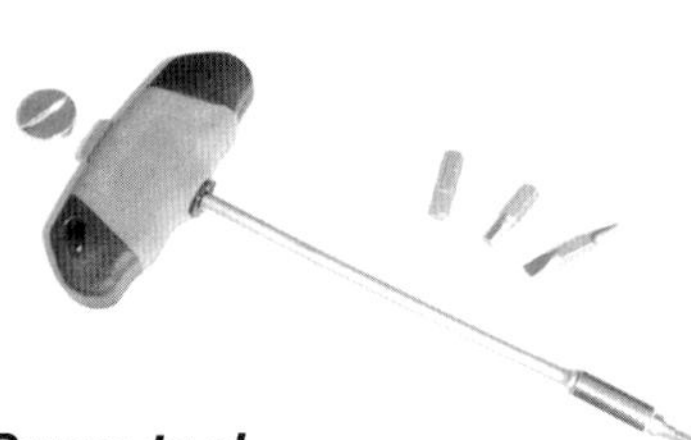

Pump tool

Circulating pumps should give many years of quiet efficient hot water circulation. They are precision manufactured to a high quality standard and are robust and resistant to corrosion. They may seize if the system water contains corrosive particles of rust and sludge.

After long periods of non-operation, rust and sludge sediment may settle within the body which cause the pump to seize when first started. Therefore, before the pump is started it is good practice to remove the vent plug and rotate the shaft using a small screwdriver or pump tool. Continue to spin the shaft until it runs freely, replace the vent plug and turn on the combi boiler.

Circulating pump overhaul

Circulating Pump Overhaul

Most modern pumps can be overhauled for cleaning etc.

1. Turn off the boiler
2. Turn off the electrical supply to the boiler and pump, remove plug(s), fuse(s) and put them in your pocket, etc
3. Remove circulating pump terminal cover
4. Caution – using a multimeter check the voltage at the pump terminals. There should be a ZERO reading
5. Drain down the central heating system or use a pipe freezer
6. Unscrew the four bolts holding the pump head to its base using a pump tool (4mm hexagonal 'T' bar)

7. Carefully pull the circulating pump head forward. Mop up water spills
8. If the water is red or black or contains debris the whole central heating system must be flushed and protected *(see Flushing and Protecting – 2.5)*
9. Carefully place a flat-head screwdriver between the bearing plate and rubber gasket. Do not try to lever off the impellor
10. Carefully lever around the circumference between the bearing plate and the rubber gasket. This will withdraw the rotor can and the rotor/impellor assembly
11. Tap the end of the rotor shaft to remove the rotor/impellor assembly from the rotor can
12. Completely remove the rotor/impellor assembly
13. Using a detergent mixture clean the rotor and shaft and the inside of the rotor can
14. Rinse thoroughly. Protect electrical components from water

Re-Assembly

1. Refit the rotor can into the motor head. Ensure the rubber upper sealing ring is in place at the top of the can
2. Fit the lower rubber gasket onto the rotor can and lightly position the rotor assembly into the rotor can
3. Re-fit the head to the pump base and hand tighten the four bolts
4. Fully tighten the four bolts in turn diagonally a little at a time, this will pull the assembly into position
5. Remove pipe freezer and allow ice plugs to melt
6. Check the rotor movement, using a pump tool or 3mm screwdriver place the tip in the end of the slot of the rotor. Make sure it moves freely
7. Replace the vent plug
8. Replace terminal cover
9. Replace any fuses and turn on the electrical supply
10. Set the programmer to central heating 'on'
11. Turn on the combi boiler
12. Reset room thermostat to original setting

6.19 Solenoids

Introduction
Solenoids are electro magnetic devices used to operate valves. A solenoid valve will control the on/off/flow of a liquid or gas through a pipe. They have been used for many years within gas appliances to control the flow of gas through the multifunctional/modureg gas valves.

Design & Construction
A solenoid valve consists of the solenoid and valve body. The solenoid consists of electrical connectors, a wire coil and a soft iron plunger often called an armature or slug. The valve body accommodates the armature, armature return spring, valve and valve seat. They are energised by either alternating current (a.c.) or direct current (d.c.). Often colour coded to indicate:-

Blue	24Vdc
Black	230Vac

Red is also used for 230Vac, so always carefully check rating.

When modulation is employed there may be 2 or 3 solenoids used *(see Gas Valves – 6.12).*

Operation
The valve is held closed by the return spring. When an electric current passes through the coil it creates a magnetic field. This will draw the armature into the solenoid, lifting the valve off its seating, allowing gas to flow. When the electric current stops, the magnetic field ceases and the return spring immediately pushes the valve onto its seating, stopping gas flow. There is often a fixed pole to assist the attraction of the armature into the coil.

The force of the magnetic field can be altered by varying the current/voltage through the coil. This modulating coil concept is used with many gas valves and these gas valves have become known as a modureg. The magnetic field is directly proportional to the current/voltage applied to the coil and therefore the movement of the armature is also proportional to the current/voltage applied. This modulation increases or decreases gas glow rate.

Many gas valves use solenoid valves that lift in two stages to give a low fire rate and a high fire rate. Alternatively, the valve may open to provide for the pilot flame and then fully open to fire the main burner.

Two solenoids may also be used within the gas valves e.g. 1 solenoid valve for the pilot flame and another to provide for the main burner.

Fault Finding
If a solenoid is suspected of being faulty or it is to be checked during fault logic procedures:-

(a) Electrically isolate combi boiler
(b) Check main terminal block connections
(c) Check valve electrical connections are clean and well fitting
(d) Check solenoid colour and read data
(e) Switch on electric supply
(f) Set multimeter to correct voltage scale and check voltage at valve electrical connections
(g) If the voltage is not correct the fault is elsewhere BEFORE the solenoid e.g. printed circuit board
(h) If the voltage is correct and the valve is not operating......
(i) Switch off electric supply and electrically isolate combi boiler
(j) Remove connectors making a note of positions
(k) Set multimeter to Ohms (Ω) test the coil resistance, if:-

Mega 0hms/overload/Infinity	=	Coil Faulty
1 Ohm or less	=	Coil Faulty

Solenoids 6.19

Solenoids (gas valves) have a range of 25V–30 V a.c. or d.c.
If 1 Ohm or less the coil has shorted out and may have blown the fuse. Check the fuse for continuity with multimeter set on Ohms (Ω), if less than 1 Ohm the fuse is O.K. If infinity is measured, replace the fuse and remember that the fuse will NOT blow again due to faulty coil.

Solenoids in a.c. circuit can make a humming sound which may be due to:-

Solenoid mountings not tight/loose – Humming sound is continual when the power is on and stops when off. Tighten and firmly fix the solenoid mountings.

Solenoid return spring loose – Humming sound is intermittent and only occurs when there is a demand for heat. The solenoid must be replaced.

Solenoid failing – Humming sound is intermittent and only occurs when there is a demand for heat. The solenoid must be replaced.

If dirt/rust finds its way inside the solenoid it can cause the movement of the armature to become erratic or the armature may completely jam. If the solenoid is installed as a separate component it may be possible to dismantle and clean the armature, the valve and valve seat. Alternatively, the obstruction may be cleared if air is blown through.

Solenoid Noise

Reason	Repair
Mountings loose	Tighten and firmly fix the mounting – *6.19*
Return spring loose	Replace solenoid – *6.19*
Solenoid failing	Replace solenoid – *6.19*

Solenoid not Energising

Reason	Repair
Poor terminal connection	Check and clean connectors. Ensure good fit – *6.19*
Burnt out coil	If resistance is zero or infinity, replace solenoid – *6.19*

Solenoid Movement Erratic

Reason	Repair
Solenoid failing	Replace solenoid – *6.19*
Armature obstruction	Clear obstruction/replace solenoid – *6.19*

Service & Repair
At service ensure terminal connectors make a good clean contact and the solenoid moves freely. Repairing solenoid valves is limited so when replacing always read data label/manufacturers instructions and replace the solenoid on a like for like basis.

6.20 Thermistors

Introduction

Domestic hot water thermistor

Thermistors are **THERM**ally sensitive res**ISTORS**. The resistance of a thermistor will change as the temperature changes. This change in resistance can be sensed at the printed circuit board. Thermistors are used within combi boilers to sense air temperature to prove the fan is working. They are more commonly used on the central heating flow pipe and domestic hot water flow pipe to monitor water temperature, often referred to as central heating and domestic hot water sensors.

In 1833 Michael Faraday first described a decrease in resistance of silver sulphide when it was heated. Over 50 years ago Sir Charles Wheatstone used this phenomenon to arrange the Wheatstone Bridge. This determines the ratio of one resistance to another when arranged in a group of 4 resistors.

Design and Construction

Versatile and rugged, thermistors are solid state, electronic devices which detect thermal changes. The small size of thermistors allows them to be positioned in awkward places and their use in combi boilers is well established due to this, their sensitivity, excellent long term stability and accuracy.

Two types of thermistors are commonly used within combi boilers:-

Positive Temperature Co-efficient (PTC) thermistors – Exhibit increasing electrical resistance with increases in temperatures and decreasing electrical resistance with decreasing temperatures.

Negative Temperatures Co-efficient (NTC) thermistors – Exhibit decreasing electrical resistance with increases in temperatures and increasing electrical resistance with decreasing temperatures.

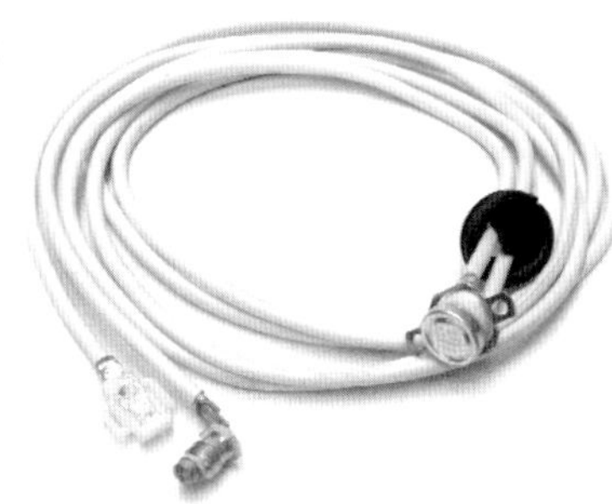

Thermistor with lead and connecting terminals

PTC thermistors are commonly manufactured from barium titanate. NTC thermistors are manufactured from the oxides of the transition metals manganese, cobalt, copper and nickel.

The NTC thermistor is most commonly used within combi boilers and most thermistors are connected to a wet pocket, requiring the combi boiler to be drained before the thermistor can be replaced. Some thermistors use dry pockets avoiding the need for draining down. In this case the pocket is packed with thermal grease to ensure effective thermal transfer.

Operation

When used to prove the fan is working effectively, usually two thermistors are used within the fan and flue assembly. These thermistors sense the difference in temperature at the fan/flue and the difference in resistance signalled at the printed circuit board. This difference will prove air movement created by the fan and therefore prove the fan is running effectively.

Thermistor

Typically, a thermistor is supplied with a small voltage, about 6V d.c. which returns to the printed circuit board either unchanged or reduced. The supplied voltage must be small to prevent the thermistor from heating itself from its own resistance, this would give incorrect readings. As the resistance of the thermistor changes the gas rate will be increased or decreased (modulate) to match the demand for heat.

So, as the temperature increases the NTC thermistor resistance decreases, the current flow increases and the potentiometer at the printed circuit board varies the voltage to the modureg.

The thermistor is one of four resistors arranged in such a way that if the resistance of one changes it is out of balance with the other three and a current will flow. This arrangement of resistors is known in electronics as a Wheatstone Bridge.

Fault Finding

A thermistor is a form of resistor and will show a resistance when tested with a multimeter set on Ohm (Ω) or Kilo-Ohm (KΩ) scale. Electrically isolate the combi boiler, disconnect the wires from the sensor to prevent other resistance from the printed circuit board being detected, check resistance across the terminals of the thermistor. If a resistance is present the thermistor is probably not faulty. But, thermistors can loose calibration and to prove correct calibration a thermistor chart must be obtained from the manufacturer. See example chart below.

To test a thermistor:-

(1) Use pipe clamp and digital thermometer to take a temperature reading from pipe as close as possible to the thermistor
(2) Disconnect wires from thermistor
(3) Using a multi meter set on Ohms (Ω) or Kilo-Ohms (KΩ) note resistance reading
(4) Compare resistance reading with manufacturers chart:-

Example of 'Negative Temperature Co-efficient' resistance chart

Temperatures (°C)

	0	**10**	**20**	**30**	**40**	**50**	**60**	**70**	**80**	**90**	**100**
0	33242	19947	12394	7947	5242	3548	2459	1740	1256	923	690
1	31534	18992	11829	7614	5036	3417	2318	1683	1217	868	671
2	29925	18088	11313	7297	4839	3291	2291	1628	1180	870	652
3	28490	17233	10813	6995	4651	3171	2212	1575	1143	845	634
4	26980	16425	10338	6707	4471	3056	2136	1524	1120	820	617
5	25633	15659	9888	6433	4300	2945	2063	1474	1075	797	600
6	24361	14934	9459	6171	4136	2840	1994	1428	1042	774	584
7	23161	14247	9052	5922	3979	2738	1926	1383	1011	752	568
8	22028	13596	8665	5685	3819	2641	1862	1339	981	730	553
9	20958	12979	8297	5458	3685	2548	1800	1297	952	710	538

Resistances (Ω)

Example:- Temperature reading of 21°C
Resistance reading of 11829 Ω (11.829kΩ)

Temperature reading of 46°C
Resistance reading of 4136 Ω (4.136kΩ)

If there is less than 1 Ohm or infinity resistance the thermistor is faulty and must be replaced. A wet pocket thermistor will require the combi-boiler/system to be drained down. A dry pocket thermistor can be replaced without draining down the combi boiler/system. Remember, pack the dry pocket with thermal grease to allow effective temperature sensing by the thermistor.

One manufacturer uses a thermistor with only one wire connection. Refer to manufacturers instructions for fault finding.

6.20 Thermistors

Thermistor Not Reading Correctly

Reason	Repair
Connections/terminals are dirty/loose	Clean connectors/terminals and ensure good fit
Open circuit/infinity caused by water ingress	Drain down combi boiler/system and replace thermistor. If dry pocket, replace thermistor and pack with thermal grease
Lost calibration	Drain down combi boiler/system and replace thermistor. If dry pocket, replace thermistor and pack with thermal grease

Service and Repair

Thermistors are robust components and require no maintenance at service. If a fault does develop they cannot be repaired or recalibrated so must be replaced. Turn off and electrically isolate the combi boiler. If the thermistor is a wet pocket type drain down the combi boiler/system or use a pipe freezer and replace faulty thermistor. If the thermistor is a dry pocket type remember to repack pocket with thermal grease.

Introduction

Also known as stats, thermostats detect a change in temperature and make or break the contacts of a switch. The switching signals the printed circuit board to sequence an event or series of events.

There are many different thermostats available, both direct and indirect acting, with even more applications. Most commonly found within combi boilers are bi-metal, liquid and vapour pressure thermostats and of course thermistors being used as:-

- Low and high limit thermostats
- Central heating overheat thermostat
- Domestic hot water overheat thermostat
- Pump over-run thermostat
- Fan over-run thermostat
- Frost thermostat
- Re-heat thermostat

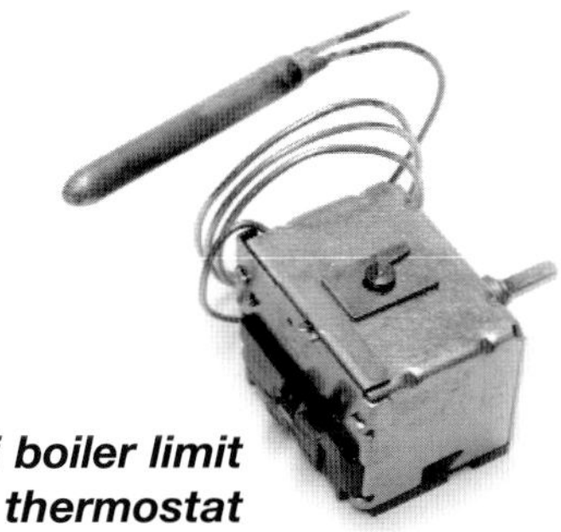

Combi boiler limit thermostat

Operation

Thermostats are thermally activated control devices used to maintain a desired temperature. The low and high limit thermostats (boiler stat) are usually either liquid/vapour expansion type indirectly operating a micro switch or bimetallic 'snap-action' type directly operating a micro-switch when satisfied or when there is a demand for heat. These thermostats interrupt solenoid valve(s) closing the gas valve or will energise solenoid valve(s) (often by signalling the printed circuit board) opening the gas valve. High limit thermostats on the central heating circuit are normally set at 83°C.

Overheat thermostats are invariably fitted to low water content boilers to protect the heat exchange from damage should the high limit thermostat fail. Again, either liquid/vapour expansion type or bimetallic snap action type as described above are used. Overheat thermostats are set to operate at between 92–98°C and when activated usually send the boiler to lock-out and must be manually reset.

Overheat thermostat

Domestic hot water overheat thermostats are self resetting normally closed type, positioned down stream of the secondary heat exchanger (calorifier). Set to operate at approximately 78°C it will move to the open position, closing the gas valve.

Domestic hot water re-heat thermostats control the gas valve to maintain the calorifier at temperature between 50–60°C.

Pump over run thermostats fitted to low water content combi boilers are again either bimetallic or liquid expansion type. In the normally open position until the temperature reaches 60°C when it will move to the closed position switching the pump to permanent live. It will remain closed keeping the pump operating after the combi boiler has shut down, dissipating the residual heat in the heat exchanger until the water temperature drops below 60°C. A built in by pass or purpose designed by pass with automatic valve is required to allow for the over run circulation.

Fault Finding

A thermostat may fail and when it does it immediately fails to supply electricity to the multifunctional control valve, often via the printed circuit board. This prevents the flow of gas and the combi boiler switches off. Alternatively, it may loose calibration and become inaccurate. In this faulty state it will turn off the gas valve at insufficient operating temperature.

When the boiler thermostat is failing or fails, the combi boiler may overheat the water. This overheat situation will cause the overheat thermostat to operate causing the boiler to shut down, often to lockout. The boiler must either be turned off, in some cases for a couple of minutes, or the reset button must be pressed before the combi boiler can be re-fired.

6.21 Thermostats

Should the overheat thermostat fail it will trip out and usually send the combi boiler to lockout. Insufficient water flow through the boiler caused by a failing/failed pump or sludge/scale build up may also cause the boiler control thermostat or the overheat thermostat to turn off or send the boiler to lockout. When using fault logic to discover the reason for the combi boiler turning off or going to lockout first establish if it is the boiler control thermostat or the overheat thermostat. Thermostat failed – To check a thermostat when cold:-

1. Turn off and electrically isolate the combi boiler
2. Using a multimeter set on Ohms (Ω) test for continuity across the terminals

Less than 1 Ohm = Thermostat O.K.

Mega Ohms/Infinity/Overload = Thermostat faulty

If the combi boiler has just turned off or just gone to lockout and the thermostat is at its trip point then:-

Mega Ohms/Infinity/Overload = Thermostat O.K.

Less than 1 Ohm = Thermostat faulty

Thermostat Lost Calibration – Using the manufacturers instructions establish set temperature and use a digital thermometer with pipe clamp or pipe sensor to establish correct calibration.

Thermostat Contact Point Worn/Burnt Out/Pitted – The classic symptom of this fault is increasing radio/TV interference. Replace the thermostat.

Thermocouple/Phial Incorrectly Positioned – The householder may have tampered with the thermostat and not correctly replaced the phial into its pocket. Check for this. When replacing a thermostat phial type ALWAYS fully insert the phial into its pocket.

Possible Causes of Faulty Thermostat

Reason	Repair
Thermostat failed	Replace thermostat – *6.21*
Thermostat lost calibration	Replace thermostat – *6.21*
Thermostat contact points worn/ burnt out/pitted	Replace thermostat – *6.21*
Thermocouple phial out of position	Reposition phial in pocket – *6.21*

Combi Boiler Shut Down (Overheat)

Reason	Repair
Expansion vessel faulty (sealed system)	Repair/replace expansion vessel – *6.7*
Air trapped in system	Vent system – *6.2*
Water leak	Find and repair leak – *6.8*
By pass valve faulty	Adjust setting or replace – *6.3*
Pump failing/failed	Repair/replace pump – *6.18*
Sludge/scale built up in heat exchanger causing kettling	Drain, flush and protect system – *2.4*

Thermistor loosing calibration or failed	Check/replace thermistor – *6.20*
Overheat thermostat faulty	Check/replace thermostat – *6.21*
High limit thermostat faulty	Check/replace thermostat – *6.21*
Gas rate too high	Establish cause and correct – *5.8/6.12*
Solenoid sticking in on position	Repair/replace solenoid or gas valve – *6.19*

Service and Repair

Thermostats are invariably factory set and if faulty are replaced (as a complete unit). They do not require servicing except to check they are positioned correctly. If the thermocouple is contained within a pocket, make sure it is inserted to the full depth and any control dial is fitted or re-fitted correctly. Check the control dial does not slip on the spindle.

Room Thermostats

Programmable room thermostat

Many manufacturers install a 24 hour mechanical time clock into the combi boiler. Used in conjunction with a room thermostat this is adequate. However, better combi boilers can use the manufacturers 7 day electronic programmer, either built in or as an optional extra, combined with a hard wired or radio frequency room thermostat, provides excellent control.

Room thermostat

The major manufacturers of programmers, room thermostats and programmable thermostats produce a wide and varied range of external controls which can be combined with most combi boilers to give the required control.

Combi boiler system wiring information

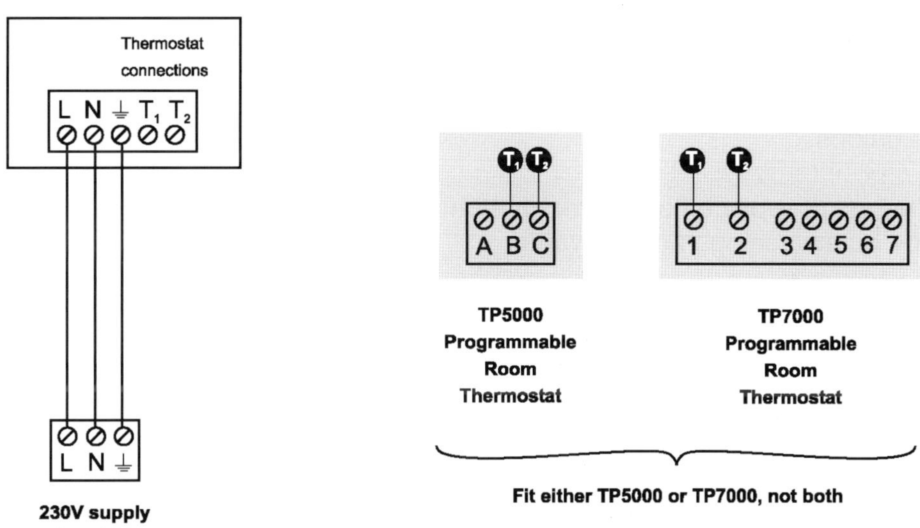

Notes

1) Refer to manufacturer's manual for wiring connections to boiler

2) A radiator thermostat should not be fitted in the room where the room thermostat is located

Frost Thermostats

Many quality combi boilers have built in frost protection. This ensures that even if the combi boiler is turned off (but electricity supply turned on) if the built in frost thermostat device detects low temperatures the combi boiler automatically switches the central heating system on and maintains a temperature high enough to prevent freezing.

Introduction

6.22 Water Governors

The rising mains cold supply to the combi boiler provides the source for the domestic hot water. Mains water pressure can vary so a water governor is installed within the appliance casing. Also known as a 'cold water inlet restrictor' or 'pressure reducing valve', the water governor regulates the flow of cold water to the heat exchanger and therefore the flow rate of domestic hot water at the draw off points.

Design and Construction

The term water governor should be taken as a generic term for the water flow control device as it takes many forms. Often it is a simple circular disc contained within the flow detector assembly which also contains the flow turbine/flow switch, possibly as a Venturi and often containing a gauze filter. It may also present itself as a reducing valve cartridge contained within a reducing valve.

Operation

Flow rates are calculated by the manufacturers to suit individual combi boilers and the water governor is factory set. Restriction of water flow to the combi boiler by using a washer or cartridge is simple but effective.

Fault Finding

The water governor may cause one fault which is typically a reduced flow rate. This must not be confused with a temporary drop in mains water supply pressure.

Reduced Domestic Hot Water Flow Rate – Due to the gauze filter/strainer being partially blocked with detritus (sand, grit, rust etc.) from the mains water supply.

Reduced Domestic Hot Water Flow Rate

Reason	Repair
Filter/strainer partially blocked by detritus	Remove and clean filter/strainer
Cartridge partially blocked by detritus	Replace cartridge
Temporary low mains supply pressure	DO NOT change flow restrictor

Service and Repair

Follow the manufacturers instructions to dismantle the water governor. Usually this involves isolating the water supply to the combi boiler, draining the domestic hot water circuit and removing governor from its unions.

Do not adjust or replace flow restrictor with any other. Always replace on like for like basis. Clean the gauze by backwashing in clean water.

Fault Finding Logic

Contents

7.0 Fault Finding Logic

Introduction

Fault finding on combination boilers is about curing the customer's problem. Their problem is the combi boiler is failing in its 'comfort' role. It is either failing in its heating role or failing in its hot water role or both. Whichever fault diagnosis procedure is used the objective is always to achieve:-

A fast and accurate diagnosis

Minimum shut down time

Effective repairs

The procedure that works for you is the best procedure but it will only achieve the objective if it includes full fault finding logic. All procedures must include a logical approach including:-

Collect Evidence – talk to the householder, attend manufacturers training courses, obtain and read the manufacturers installation and servicing instructions, talk to colleagues and component wholesaler staff, keep a written record of combi boilers worked on, fault, repair, cost of component(s), time taken.

Analyse Evidence – Use knowledge of components installed, knowledge of operating principles, knowledge of firing sequence.

Locate the Fault – Use fault logic, use safe working practices, use instruments to test signals/components, prove fault, avoid wrong conclusions.

Establish Cause of Fault – Use knowledge of components and operating principles to remove cause of fault.

Repair the Fault – Use professional tools, use quality parts a from a reputable wholesaler, avoid damge to other components, use pipe freezing machine if drain down necessary.

Check the System – Test all external and internal time and temperature controls, observe firing and shutting down sequence in both central heating and domestic hot water modes. Check central heating and domestic hot water temperatures, check flow rates, check gas rates and carry out tightness test if appropriate, check correct combustion with a flue gas analyser.

Follow safe working procedures

Carry out electrical safety checks

Much has been written concerning fault finding methods and procedures including, sequential, halving or sub division, bracketing etc. all equally good. They all call for a methodical approach to the diagnostic process. Follow the procedure that works for you so that you have a logical and efficient method of diagnosing combi boiler faults.

Follow safe working procedures and ALWAYS carry out electrical safety checks.

Fault finding logic does get easier with practice and as your approach develops it is always rewarding to fix something that is broken so:-

Have fun!

Introduction

The pressure gauge will indicate the system pressure when cold through to hot. A discharging pressure relief valve and frequent use of the filling loop indicate a fault or fault developing. The pressure will indicate the system pressure at cold and through to hot.

Use the pressure gauge and temperature gauge reading in conjunction with pressure relief valve discharge frequency and filling loop use to establish a fault find logic procedure.

Temperature gauge	Pressure gauge	
Normal	High	Fault is with expansion vessel/valve/ connecting pipe/domestic hot water heat exchanger/filling loop
High	High	Fault is with limit thermostat/pump/ scaling

Regular Topping Up of System

Reason	Repair
Excessive venting of radiators by householder	Establish reason for air in system, trace reason, correct and re-pressurise system – *see 'Central Heating – Fault Finding & Repair' book*
Leak in system at pump or radiators unions, pipe work, heat exchanger cracked, etc	Raise system pressure to 3 bar using the filling loop, trace and repair leak, release water until pressure is at 1 bar – *6.17*

Regular Topping Up of System and Pressure Relief Valve Discharges Pressure Gauge Low/Zero When Cold

Reason	Repair
Expansion vessel lost charge naturally over time	Drain down, re-pressurise vessel and top up system – *6.7*
Expansion vessel diaphragm split or porous	Check for water at schraeder valve when depressed. Replace expansion vessel – *6.7*
Schraeder valve faulty and letting out nitrogen/air	Check for air leak at valve with LDF. Re-tighten/replace valve if faulty – *6.7*
Pressure relief valve faulty/leaking due to obstruction at seating	Clear obstruction or replace pressure relief valve. Raise to 3 bar, operate pressure relief valve and reduce to 1 bar – *6.7*
Pressure over 3 bar, water discharge at pressure relief valve – expansion vessel undersized	Replace with larger or fit additional vessel – *6.7*
Vessel to pipe work connecting pipe blocked with sludge	Remove vessel and clear blockage or replace connecting pipe. Refit vessel – *6.7*
Pressure gauge reading low. Pressure relief valve discharges	Check/replace faulty pressure gauge and sensing head – *6.17*
Excessive fill pressure	Reduce pressure to 1 bar (cold) by releasing water from system – *6.8*

Regular Topping Up of System/Pressure relief Valve Discharges/System Overheating

Reason	Repair
Combi boiler overheating due to scaling or sludge and pressure relief valve discharging	Drain, flush and protect central heating system – *2.5*
Combi boiler overheating due to pump failing/failed and pressure relief valve discharging	Repair/replace pump – *6.18*
Limit thermostat faulty so overheating to overheat thermostat and pressure relief valve discharging	Check/replace limit thermostat – *6.21*
Pump or fan over run faulty	Check pump – *6.18* Check fan – *6.11*
Manual or Automatic By-pass set incorrectly	Reset/replace By-pass valve – *6.3*

System Cold, Pressure Relief Valve Discharges and High Pressure at Gauge (3bar)

Reason	Repair
Domestic hot water heat exchanger leaking across to primary system	Replace domestic hot water heat exchanger – *6.13*
Filling loop attached and isolators letting by	Fully tighten isolators, remove filling loop and inform customer – *6.8*

Introduction

When the pressure relief valve discharges always check the filling loop is not attached and letting by. Also check the temperature and pressure gauges with the system cold and hot.

Check the pump is running, smoothly and is not noisy. The fault may be with the expansion vessel, vessel valve or connecting pipe. If the domestic hot water heat exchanger is not leaking across to the primary system the thermostats may be faulty or gas rate too high.

Valve Discharge at Fill-Up

Reason	Repair
System pressure greater than 3 bar	Remove filling loop and reduce system pressure to 1 bar (cold) – *6.8*
Filling loop attached and letting by	Fully close isolators or replace and remove filling loop – *6.8*
Pressure gauge incorrectly reading less than 3 bar	Replace pressure gauge (like for like) – *6.17*
Pressure relief valve faulty (unlikely)	Replace pressure relief valve – *6.16*

Valve Discharge

Reason	Repair
Pump failing/failed, causing water in heat exchanger to boil	Check and repair/replace pump – *6.18*
Pump over run failing/failed	Trace and repair fault at over run thermostat/ thermistor or at printed circuit board – *6.18*
Expansion vessel under pressurised	Re-pressurise expansion vessel – *6.7*
Schreider valve leaking	Tighten or replace Schreider valve and re-pressurise or replace expansion vessel – *6.7*
Expansion vessel diaphragm split or porous	Check for water at schraeder valve when depressed. Replace expansion vessel – *6.7*
Expansion vessel under sized for installation	Replace with correct size vessel or fit additional expansion vessel – *6.7*
Gas rate too high	Check injector and measure gas rate/adjust burner pressures – *5.8*
Limit/overheat thermostat faulty	Replace thermostat – *6.21*
Obstruction preventing valve closing on seating	Clear obstruction or replace pressure relief valve – *6.16*
Domestic hot water heat exchanger leaking across to primary system	Replace domestic hot water heat exchanger – *6.13*
Attached filling loop letting by	Fully close isolators, remove filling loop and inform customer – *6.8*

7.3 Combi Boiler Noise

Introduction

Modern combi boilers operate very quietly so any significant noise is a sign of a fault or fault developing. Identify where the noise is originating from and correct. Remember, when a pressure relief valves activate it tends to cycle between open and closed, causing a banging sound. Again this sound is a sign of a fault or fault developing.

Fan Runs Noisily

Reason	Repair
Fan unbalanced	Clean away dirt and fluff – *6.11*
Bearings wearing out	Lubricate bearings/replace fan – *6.11*
Fan slipping on shaft	Tighten grub screw – *6.11*

Solenoid Noise

Reason	Repair
Mounting loose	Tighten and firmly fix – *6.19*
Return spring loose	Replace solenoid – *6.19*
Solenoid failing	Replace solenoid – *6.19*

Heat Exchanger Noise

Reason	Repair
Kettling due to sludge/scale build up	Drain, flush and protect – *2.5*
Pump failing causing water to overheat	Service/replace pump/pump head – *6.18*
Pump/fan over run faulty causing water to overheat	Check pump/fan over run – *6.18*
Limit thermostat failed so water overheated to overheat thermostat	Check/replace limit thermostat – *6.21*
By-pass valve incorrectly set	Re-set or replace By-pass valve – *6.3*

Explosive Ignition

Reason	Repair
Armature in gas valve sticking	Check solenoids/clear obstruction or replace solenoids/gas valve – *6.12*
Burner injector oversize	Check injector condition/size, clean or replace – *5.8*
Gas rate too high	Check gas pressures – *5.8*
Pilot burner assembly incorrectly positioned	Re-position pilot assembly – *6.14*

Pump Noise

Reason	Repair
Air in system/pump	Vent system and pump – *6.18*
Dry running of pump	Position combi boiler correctly
Impellor out of balance due to wear and tear	Replace pump/pump head – *6.18*

Introduction

Not all combi boilers have a pilot flame. However, many still do and faults can cause the firing sequence to fail at the pilot flame stage. If the fan is running and there is no pilot flame check fan proving at air pressure switch and printed circuit board. Check signal from printed circuit board to the gas valve and check solenoid. Finally check the ignition sequence, spark quality and the flame safety device. If the fan is not running the fault will be at or before the fan.

Pilot Flame Too Big/Lifting Off

Reason	Repair
Excessive gas pressure	Adjust working gas pressure – *5.8*
Over sized pilot injector	Check and replace pilot injector – *6.14*
Incorrect adjustment	Check and adjust pilot injector – *6.14*
Fan failing	Check fan speed/operation and flue – *6.11*
Air pressure switch/air flow or hoses, thermistor faulty	Check air pressure switch and hoses or test flue thermistor – *6.1*

Pilot Flame Too Short

Reason	Repair
Insufficient gas pressure	Adjust working gas pressure – *5.8*
Partially blocked pilot injector	Clean pilot injector – *6.14*
Incorrect adjustment	Check and adjust pilot injector – *6.14*
Partially blocked pilot gas filter	Clean pilot gas filter – *6.14*
Partially blocked pilot supply tubes	Clean pilot supply tubes – *6.14*

Pilot Flame Extinguishing

Reason	Repair
Excessive draughts	Check flue and flue terminal are correct – *6.11*
Vitiated Air	Check flue and flue terminal for obstruction – *6.11*
Explosive ignition	Check armature in gas valve not stuck – *6.12*
Pilot flame incorrectly located	Correctly re-position pilot assembly – *6.14*
Pilot flame too short	Adjust working gas pressure – *5.8*
Fan failing	Check fan speed/operation and flue – *6.11*

Introduction

When the central heating is O.K. but the domestic hot water is not O.K. fault logic eliminates all operating sequence possible faults such as electric supply, fan, pump ignition/pilot sequence, flame proving device, gas rate, gas valve and solenoids, burner pressure, central heating thermostats and thermistors, heating side of printed circuit board etc.

Therefore, the fault can only be at the diverter valve, micro switches, flow switch, domestic hot water thermostats/thermistors, domestic hot water exchanger, water governor, printed circuit board etc.

No Domestic Hot Water
Diverter Valve Diaphragm Pass Type

Reason	Repair
Split diaphragm	Replace diaphragm – *6.5*

Diverter Valve Venturi Type

Reason	Repair
Split diaphragm	Replace diaphragm – *6.5*
Obstruction in Venturi i.e. scale, rust etc	Remove diverter valve/clean and blow through passage way – *6.5*

No Domestic Hot Water or Domestic Hot Water not at Temperature and Central Heating Circuit Warms
Diverter Valve

Reason	Repair
Wax lost calibration	Drain down boiler and replace capsule – *6.5*
Capsule scaled	Drain down boiler and replace capsule – *6.5*

Spindle Does Not Move

Reason	Repair
Pump failing/failed	Repair/replace pump – *6.18*
Venturi pipe work blocked	Clean and blow through Venturi pipe work – *6.5*
Filter/strainer blocked	Clean strainer. Check water governor – *6.22*
Diaphragm perished/split	Replace diaphragm or diverter valve – *6.5*

Slow Movement of Spindle

Reason	Repair
Pump failing	Repair/replace pump – 6.18
Spindle shaft scaled due to leak	Drain down boiler. Clean spindle or fit new spindle and sealing unit – *6.5*
Venturi partially blocked due to rust or sediment	Drain down boiler, dismantle valve, clean and blow through Venturi tube – *6.5*
Water flow slow due to sealed domestic hot water heat exchanger	Drain down boiler. Replace domestic hot water heat exchanger – *6.13*
Mains pressure low	Investigate reasons. Check water governor – *6.22*
Pressure loss at spindle sealing unit	Repair or replace unit or valve – *6.5*
Air may be trapped in valve after repair	Slacken unions to vent air, retighten – *6.5*

Spindle Moving but no Response

Reason	Repair
Not enough spindle force due to faulty diaphragm causing pressure loss	Drain boiler/replace diaphragm – *6.5*
Not enough spindle force due to sealing unit leaking	Drain boiler/replace sealing unit – 6.5
Micro-switch faulty	Replace micro-switch (flow switch) – *6.5*
Printed Circuit Board not sequencing, fan not working, ignition sequence (sparking) or faulty gas valve etc	Prove micro-switch is not faulty and trace fault to next possible component – *6.6*

No Hot Water/Central Heating O.K.

Reason	Repair
Thermistor connections/terminals are dirty/loose	Clean connectors/terminals and ensure good fit – *6.20*
Open circuit/infinity caused by water ingress	Drain down combi boiler/system and replace thermistor. If dry pocket, replace thermister and pack with thermal grease – *6.20*
Thermistor – lost calibration	Drain down combi boiler/system and replace thermistor. If dry pocket, replace thermister and pack with thermal grease – *6.20*
Thermostat failed	Replace thermostat – *6.21*
Thermostat lost calibration	Replace thermostat – *6.21*
Thermostat contact points worn/ burnt out/pitted	Replace thermostat – *6.21*
Thermocouple phial out of position	Reposition phial in pocket – *6.21*

Domestic Hot Water Not at Temperature

Reason	Repair
Sludge/scale build up in main heat exchanger often causing noise – kettling	Drain, flush and protect – *2.5*
Sludge/scale or detritous in domestic hot water heat exchanger	Replace domestic hot water heat exchanger – *6.13*

Hot Water Alternating Hot and Cold

Reason	Repair
Sludge/scale build up in domestic hot water heat exchanger	Drain, flush and protect system or replace domestic hot water heat exchanger – *2.5*

Reduced Domestic Hot Water Flow Rate

Reason	Repair
Filter/strainer partially blocked by detritus	Remove and clean filter/strainer – *6.22*
Cartridge partially blocked by detritus	Replace cartridge – *6.22*
Temporary low mains supply pressure	DO NOT change flow restrictor

7.6 Central Heating

Introduction

When the domestic hot water is O.K. but the central heating is not O.K. fault logic eliminates all operating sequence possible faults such as electric supply, fan, pump, ignition/pilot sequence, flame proving device, gas rate, gas valve and solenoids, burner pressure, domestic hot water thermostats and thermistors, hot water side of printed circuit board etc. Therefore, the fault can only be at the diverter valve, micro switches, flow switch, central heating thermostat and thermistors, printed circuit board etc.

External controls such as the room thermostat or programmable room thermostat must be calling for heat. Without a heat demand the combi boiler will not sequence.

Central Heating Not at Temperature

Reason	Repair
Sludge/scale build up in main heat exchanger often causing kettling	Drain, flush and protect – *2.5*
Pump speed slow due to sediment/sludge etc	Flush and protect system and overhaul pump – *2.5*

No Central Heating/Hot Water O.K.

Reason	Repair
Room thermostat faulty so no signal to combi boiler	Check call for heat at room thermostat *– see 'Central Heating – Fault Finding & Repair' book*
Diverter valve faulty	Check diverter valve – *6.5*
Dripping hot water draw off point preventing diverter valve moving	Fully close hot water draw off point
Micro switches faulty	Check micro switch – *6.5*
Thermistor connections/terminals are dirty/loose	Clean connectors/terminals and ensure good fit – *6.20*
Open circuit/infinity caused by water ingress	Drain down combi boiler/system and replace thermistor. If dry pocket, replace thermistor and pack with thermal grease – *6.20*
Thermistor lost calibration	Drain down combi boiler/system and replace thermistor. If dry pocket, replace thermistor and pack with thermal grease – *6.20*
Thermostat failed	Replace thermostat – *6.21*
Thermostat lost calibration	Replace thermostat – *6.21*
Thermostat contact points worn/ burnt out/pitted	Replace thermostat – *6.21*
Thermocouple phial out of position	Reposition phial in pocket – *6.21*

Introduction

If the fan is running at speed any faults must be after the fan so the following will not be faulty, electrical supply (check polarity), fuse(s), heat demands, external heating controls, pump, limit and overheat thermostats have been proved, diverter valve, micro switches, power side of printed circuit board etc. With the fault at or after the fan the following must be checked, air pressure ports and hoses, air pressure switch, micro switch or flue thermistor, combustion chamber and case seals, ignition system, re-ignition side of printed circuit board etc.

Fan Running O.K./No Ignition Sparking

Reason	Repair
Fan electrical connectors loose	Firmly attach terminals and ensure good contact – *6.11*
Air pressure hoses loose	Firmly fix air pressure hoses to ports – *6.1*
Air pressure hoses split/perished	Replace air pressure hoses – *6.1*
Condensation inside air pressure hoses	Remove, blow through or replace – *6.1*
Brass insert at port is blocked/partially blocked	Remove hose and blow through brass insert – *6.1*
Air pressure hoses +ve/-ve pressure connections reversed	Correctly re-position hoses – *6.1*
Air pressure switch diaphragm damaged/perforated	Replace Air Pressure Switch – *6.1*
Electrical micro-switch faulty	Replace Air Pressure Switch – *6.1*
Combustion chamber case seals in poor condition	Replace seals – *6.11*
Pilot window glass broken	Replace glass – *6.11*
Pilot window seals in poor condition	Replace glass seal – *6.11*

Fan Runs Slowly/No Ignition Sparking

Reason	Repair
Fan unbalanced	Clean away dirt and fluff – *6.11*
Bearings wearing out	Lubricate bearings/replace fan – *6.11*
Fan slipping on shaft	Tighten grub screw – *6.11*

7.7 No Sparking

Piezo Electric Ignitors - No Spark

Reason	Repair
Crystals damp	Replace ignition capsule/device – *6.14*
Crystals breaking down	Replace ignition capsule/device – *6.14*
Electrode lead terminal connection(s) unsound or disconnected	Clean/repair/remake connection – *6.14*
Electrode lead insulation damaged	Replace ignition electrode lead – *6.14*
Ignition electrode incorrectly positioned	Reposition ignition electrode – *6.14*
Ignition electrode tip pitted/damaged	Replace ignition electrode – *6.14*
Poor earth preventing complete circuit	Establish good earth – *6.14*

Introduction

If sparking is occurring we know the following are working correctly:-
The air pressure switch/thermistor, air pressure hoses, fan combustion and case seals, fan, diverter valves, limit and overheat thermostats, pump, heat demand i.e. external heating controls, fuses, electrical supply etc.

Therefore the fault must be at the pilot flame if there is one, or at the ignition system assembly including wires and connections, gas valve, solenoids, printed circuit board etc.

Gas Valve not Energising – Flame Supervision Devices 1

Reason	Repair
Tip not correctly positioned in pilot flame	Check pilot flame is correct and/or reposition tip – *6.14*
Pilot flame too short	Clean pilot assembly – *6.14*
Pilot flame misaligned	Clean and/or realign – *6.14*
Tip burnt out	Replace thermocouple – *6.9*
Build up of soot/dirt on tip	Carefully clean tip – *6.9*
Insulation damaged causing short	Check resistance, high or infinity, faulty, replace thermocouple – *6.9*
Connection to gas valve loose	Remake connection – *6.12*
Electro-magnetic valve faulty	Check gas valve – *6.12*

Gas Valve not Energising – Flame Supervision Devices 2

Reason	Repair
Probe not correctly positioned	Check position – 6.9
Rectification lead faulty	Check continuity and replace if faulty – *6.9*
Faulty connections	Clean and remake connections – *6.9*
Dirt blocking burner close to rectification probe	Clean burner of fluff and dirt – *6.9*
Reversed Polarity	Check electric supply – *3.4*
Electromagnetic valve faulty	Check gas valve – *6.12*

Solenoid not Energising

Reason	Repair
Poor terminal connection	Check and clean connectors. Ensure good fit – *6.19*
Burnt out coil	If resistance is zero or infinity, replace solenoid – *6.19*

7.8 Sparking – No Flame

Solenoid Movement Erratic

Reason	Repair
Solenoid failing	Replace solenoid – *6.19*
Armature obstruction	Clear obstruction/replace solenoid – *6.19*

Main Burner Will Not Fire

Reason	Repair
Gas supply/pressure suspect	Check standing gas pressure – *5.8*
Burner injector blocked	Check and clear blockage – *5.8*

Introduction
With no central heating and no hot water remember to check:-

1. The electrical supply to the combi boiler
2. Fuse(s) in connection unit, plug and with in the combi boiler
3. Lockout button and reset if required
4. External and internal time and temperature controls are set to 'call' for heat
5. First thermostat if fitted
6. Gas supply pressure
7. Check the pump is running at speed.

If the fan is running, the fault is at or after the fan. If the ignition system is sparking the fault is at or after the ignition assembly etc.

No Central Heating – No Hot Water

Reason	Repair
Domestic hot water switch faulty	Set demand and check continuity. Less than 1 ohm O.K. or replace – *6.20*
Diverter valve micro switch faulty	Check/replace micro switch – *6.5*
Pump failing/failed	Check pump, repair/replace – *6.18*
Flow switch faulty	Check/replace flow switch – *6.10*
No power from printed circuit board	Check for 230V at fan terminals – *6.11*
Air pressure failed at normally open	Check air pressure switch repair or replace – *6.1*
Fan motor winding burnt out	Check resistance or coil. If faulty replace fan – *6.11*
Bearings wearing out	Check balance and feel/Replace fan – *6.11*
Rotor jammed	Free rotor if possible or replace fan – *6.11*
Low system pressure, lost through time	Re-pressurise system to 1 bar – *6.8*
Faulty low pressure switch	Replace switch or use blanking kit (check pressure gauge is not faulty – *6.15*)
Pressure loss due to leak	Repair leak – pressurise system to 1 bar – *6.15*
Pressure loss due to overheat and water discharge through pressure relief valve	Establish cause of overheat – *6.16*

Fan Runs Slowly

Reason	Repair
Fan unbalanced	Clean away dirt and fluff – *6.11*
Bearings wearing	Lubricate bearings/replace fan – *6.11*
Fan slipping on shaft	Tighten grub screw – *6.11*
Pressure loss due to undersize/low charge/ faulty expansion vessel	Recharge/replace vessel or fit additional expansion vessel – *6.10*
Air in system	Vent system at automatic/manual air vent(s), radiator vents and at pump vent screw – *6.2*
Pump failure – no electrical supply	Check electrical connections at pump – *6.18*
Pump failure – burnt out motor windings	Replace pump head – *6.18*
Pump failure – rotor seized due to rust/sludge sediment	Repair impellor movement or replace pump or pump head – *6.18*
Pump failure – worn out bearing(s) due to wear and tear	Replace pump head – *6.18*
Pump failed – dry running	Replace pump head and vent before turning on combi boiler – *6.18*
Pump failure – burnt out motor windings due to seized rotor	Replace pump head – *6.18*
Voltage is leaving printed circuit board	Check and replace. Caution – DO NOT jump to conclusions about printed circuit board – *6.6*
High limit boiler thermostats/thermistor faulty	Check and replace – *6.20/6.21*
Overheat thermostat faulty	Check and replace – *6.20*
Hot water flow switch stuck in demand position	Check and replace – *6.10*
Air pressure micro switch faulty	Check switch is in 'no air' position – *6.1*

Introduction

If the combi boiler is operating correctly and either shuts down or goes to lockout the controls are responding to a fault. The fault must be with either air in the system causing the low pressure switch to signal the printed circuit board, a faulty low pressure switch, faults with the automatic air vent, the heat exchanger contents boiling, boiler thermostat faulty, gas rate, gas valve etc.

Combi Boiler Shutting Down Irregularly

Reason	Repair
Vent cap closed	Manually vent system and open vent cap several turns – *6.2*
Seized float/needle valve due to rust/ sludge/scale	Remove and clean or replace vent – *6.2*

Boiler Cycling

Reason	Repair
Sludge/scale build up	Drain, flush and protect – *2.5*
Boiler thermostat failed	Replace boiler thermostat – *6.21*
No means of providing boiler interlock	Fit room/programmable thermostat – *6.21*

Combi Boiler Shuts Down/Lockout Before 'Demand' is Satisfied

Reason	Repair
Additional radiators fitted with thermostatic radiator valves	Increase automatic By-pass valve setting – *6.3*
Additional thermostatic radiator valves added	Increase automatic By-pass valve setting – *6.3*
Last/highest radiator not reaching temperature	Increase automatic By-pass valve setting and/or increase lockshield valve opening – *see 'Central Heating – Fault Finding & Repair' book*
No automatic By-pass valve fitted	Fit automatic By-pass valve and set correctly – *6.3*
Room/Programmable room thermostat set too low	Re-set room/programmable thermostat – *6.21*

Boiler Lockout

Reason	Repair
Condensate formed ice plug in pipe work	Remove ice plug (use warm water), insulate pipe work – *6.4*
Condensate not running away	Adjust condensate pipe work to steeper incline – *6.4*

Combi Boiler Fires O.K. when Demand Stops, Goes to Lockout

Reason	Repair
Pump or fan over-run has failed, overheat thermostat operated, combi goes to lockout	Check fan and/or pump O.K. also check overheat thermostat

Combi Boiler Shut Down (Overheat)

Reason	Repair
Expansion vessel faulty (sealed system)	Repair/replace expansion vessel – *6.7*
Air trapped in system	Vent system – *6.2*
Water leak	Find and repair leak – *6.8*
By pass valve faulty	Adjust setting or replace – *6.3*
Pump failing/failed	Repair/replace pump – *6.18*
Sludge/scale built up in heat exchanger causing kettling	Drain, flush and protect system – *2.4*
Thermistor loosing calibration or failed	Check/replace thermistor – *6.20*
Overheat thermostat faulty	Check/replace thermostat – *6.21*
High limit thermostat faulty	Check/replace thermostat – *6.21*
Thermostat phial incorrectly positioned	Re-position phial – *5.8*
Gas rate too high	Establish cause and correct – *5.8/6.12*
Solenoid sticking in on position	Repair/replace solenoid or gas valve – *6.19*
Gas regulator at gas meter faulty	Check standing/working pressures, report fault to gas supplier – *2.1*
Undersize gas supply pipework	Check working pressure with other appliances turned on. Replace undersize pipework – *2.1*
Flame supervision device incorrectly positioned/faulty	Check FSD position, re-position or replace if faulty – *6.9*

Index

Abbreviations

A	Ampere
AC	Alternating Current
ACS	Accredited Certification Scheme
AMP	Ampere
ASD	Atmosphere Sensing Device
AV	Automatic Vent
C	Centigrade
CH4	Methane (Natural Gas)
CO	Carbon Monoxide
CO2	Carbon Dioxide
Combi	Combination Boiler
CORGI	The Council for Registered Gas Installer
D.C.	Direct Current
DHW	Domestic Hot Water
DSI	Direct Spark Ignition
F.F.	Fan flued
F.F.D.	Flame Failing Device
F.S.D.	Flame Supervision Device
H.S.E.	Heath & Safety Executive
IP	Intermittent Pilot
L.D.F.	Leak Detection Fluid
L.P.G.	Liquid Petroleum Gas (Propane)
M	Metres
mm	millimetres
MCV	Meter Control Valve
M.O.P.	Maximum Operating Pressure
N.T.C.	Negative Temperature Co-efficient
pH	'power' hydrogen – acidity/alkalinity
P.R.V.	Pressure Relief Valve
P.T.V.	Positive Temperature Co-efficient
p.p.m.	Parts per Million
P.V.C.	Polyvinylchloride
R.S.	Room sealed
SEDBUK	Seasonal Efficiency of Domestic Boilers in the United Kingdom
Stat	Thermostat
TTB	Themishe Terugslag Beveiliging (Dutch) monitoring device spillage
U.V.	Unvented
V	Volts
W	Watts